MAN IN A SPIDER-NET

MAN IN A SPIDER-NET

SVEN RASEGÅRD

Nova Science Publishers, Inc.
New York

For permission to use material from this book please contact us:
Telephone 631-231-7269; Fax 631-231-8175
Web Site: http://www.novapublishers.com

NOTICE TO THE READER

LIBRARY OF CONGRESS CATALOGING-IN-PUBLICATION DATA
Rasegård, Sven.
Man in a spider-net / Sven Rasegård.
p. cm.
Includes index.
ISBN 1-60021-414-2
1. Social ecology. 2. Science--Social aspects. 3. Technology--Social aspects. 4. Communities. 5. Ecology. I. Title.
HM861.R37 2006
304.201--dc22 *2006025941*

Published by Nova Science Publishers, Inc. ≃ New York

CONTENTS

DEDICATION

To Elsie,
My Wife and Best Friend

PREFACE

"The golden rule is that there are no golden rules".
George Bernhard Shaw

This report is mainly based on the author´s earlier published papers in certain scientific journals and can be seen as an attempt to give a glimpse of organizational systems from micro to macro. But why has the name of the report been chosen? The author´s answer is that the report is an attempt to specifically point out how human life is influenced by systems, paradigms and social conventions. As a consequence, human beings can be considered trapped in a spider net of the mentioned factors from cradle to grave. This answer, however, seems to not be enough to motivate me to begin this report with a description of an example of organizational systems. So why begin this way? The answer is that man in our modern world cannot without difficulties escape from organizational systems - if he does not prefer life on an isolated island like Robinson Crusoe. They infiltrate the everyday life of human individuals . Rational structured organizations, namely, are prerequisites for the society´s deliveries of products and services of good quality at reasonable prices from organizations of different kinds, public as well as others. Not the least of which are scientific organizations such as universities and related institutions.

Thus, what has been said above seems enough motivation to begin the first part of this report with a study of organizational structuring. It can be added that human individuals who are employed in different organizations depend on the organizational structure and that a good or a bad structure of an organization can influence their life as employees and even influence their private life. In order to form a model of an organization connected to certain ideas of cybernetics, 20 critical subsystems have been chosen. These subsystems have been created by the American Professor James Grier Miller. However, it is not enough to form a model for organizational structuring . It is also of utter importance to couple this structure to a structure of an economic system. In this case, the zero-base budgeting created by Peter A. Phyrr, when he was employed at Texas Instruments, comes into focus as a tool for budgeting and planning.

However, the organizational systems are only parts of superior systems that dominate the society. These systems are intertwined with paradigms and social conventions. It is these systems which are the objects of a study in part two. Thus, it is possible to say that the systems in question directly influence human life in both social and cultural ways as well as influencing scientific research. In order to set the systems in a frame of human life, certain central thoughts of Karl Popper are expressed. These thoughts are followed with an overview of functions of a politically democratic system and its malfunctions, especially showing that an originally democratic system can develop into a corporate system. The danger of corruption and indoctrination by ideas of a certain political color in order to catch citizens into a political pattern loyal to the ruling political paradigm will also come into focus. This can also be called a form of brain washing.

In this connection of systems and paradigms, it is utterly important that the scientific research can be done without unnecessary restrictions of certain paradigms as well as restrictions of a certain political system. Here the danger can be that research not conformed to a certain political idea can be denied necessary grants from the State and its closely related institutions. When it comes to paradigms, certain research which seems to break an established paradigm can be banned. Scientists can as a consequence be denied the publication of their research results in influential scientific journals. And if they persist holding their results as true, they can risk a ruined career. A serious danger in this connection is that politicians entering the boards of universities and scientific institutions let their political color dye the objects of research in order to conform these or a part of them to the respective party´s ideological pattern. At last an overview of the behavior of human individuals and their connections to the social life come into focus.

INTRODUCTION

This report can be regarded as an attempt to set focus on certain central problems of human society and is divided into two main parts. Part one is oriented towards organizational systems and part two can be regarded as a study of how systems and paradigms influence human life.

Part One

This part gives an overview of man´s relations to organizational systems using systems science theory because a society, as known, is built on different organizations. As an instrument, an observational/system has been chosen. This system consists of 20 critical subsystems created by the American Professor James Grier Miller. As a model for application of these subsystems to an organization, a scientific community has been selected. In this connection, these critical subsystems have been used as a diagnostic instrument forming a base for investigations of malfunctions and proposals for improvement. However, it is not enough to investigate an organization´s functioning with proposals for improvements. An utterly important question is also to investigate the economic structure of the organization and in this connection take into consideration the goal-setting on different levels of the organization as a base for judging the efficiency and effectiveness. In other words, to determine how the organization is able to achieve its goals both internally and externally. How the organization can be considered to carry out its activities concerning goal-setting will be based on an analytical description coupled to a learning function or, in other words, which types of mistakes have been made and vise versa what type of factors has lead toward a positive development. The viewpoints of the system's impact and its consequences are indicated in a summary which ends this part of the report.

Part Two

In this part of the report which deals with the paradigmic and social impact on human individuals, certain central thoughts of Karl Popper are accounted for. Also in this part the paradigmic influence on scientific researchers is set into focus. Certain scientists who give life to ideas that do not correspond to the present established paradigm, namely, risk to be out-frozen and their carrier can be ruined . To go back to the systems in part one, it can be said that the human individual can be entangled in systems of a kind other than what has been shown in part one – political and social. In this connection, it can be in place to scrutinize the real function of the Western Democracy and certain not so good features of this boasted form of political government. And so we come to human behavior. Here the author has made an attempt to show a structuring of certain main concepts connected with examples. This is done because, as known, the various kinds of behaviors are shaping the human life in society both in good and bad aspects. This part of the report ends with a summary pointing out certain central dangerous phenomena in our western civilization.

Addition

In a special addition to this report an example of a case concerning a malfunction of an investigation from the practical world is given. This case is written in the form of a fairy tale. However, the case is based on real facts.

PART ONE

THE SYSTEM CONCEPT AND APPLIED SYSTEMS

Chapter 1

An Outline of the System Concept

1.1. Some Words of Introduction

A prerequisite for a system to fill a useful function in improving conditions on different levels in the world of mankind is that it is applicable. A pure philosophical system can not be applied without adjustments to the demands of the practical world. However, the philosophical systems fill the function of a lighthouse guide to the constructors of applicable systems in order to guide the applicable systems into a right route. In this connection the General Systems Theory created by the German-Canadian Professor Ludwig von Bertalanffy contains the mentioned function. It outlines the overlapping thoughts about the system problems, but not in the form of directly applicable systems. Thus, an overlapping theory is necessary to serve as a guiding lighthouse for further direction of the system work of constructing applicable systems. Bertalanffy says in this context: "One would do well to remember the old Kantian maxim that experience without theory is blind, but theory without experience is a mere intellectual play." (Bertalanffy, 1968, p. 101). But, what does this mean in regards to a system? The answer can be rather complicated. However, the author will make an attempt to clarify the fundamental criteria of the system concept further in this paper – even if a wide circle of researchers of systems science are familiar with certain aspects of systems.

However, we can conclude that there are various types of systems. The author, nevertheless, has selected two types of systems for closer examination. Comments on these systems have been published earlier in certain scientific journals and, perhaps, have been read by some researchers of systems science. The reason for this re-publishing is that the author considers the described types of systems of great importance in relation to other systems within the frame of man in a web of systems. These systems can both be classified as artifactual- respective observational/analytical systems.

Maybe, this report can help to shed light on certain points which have not been brought into focus earlier, even if certain researchers will denote some of what is said here as banalities. Nevertheless, the author is willing to take this risk.

1.2. Certain Characteristics of Systems

To begin it can be of interest to give the following definition: "A system is a set of at least two elements and a relation that holds between each of its elements and at least another element in the set." (Clemson, 1984, p. 250). This definition can be illustrated by Figure 1.

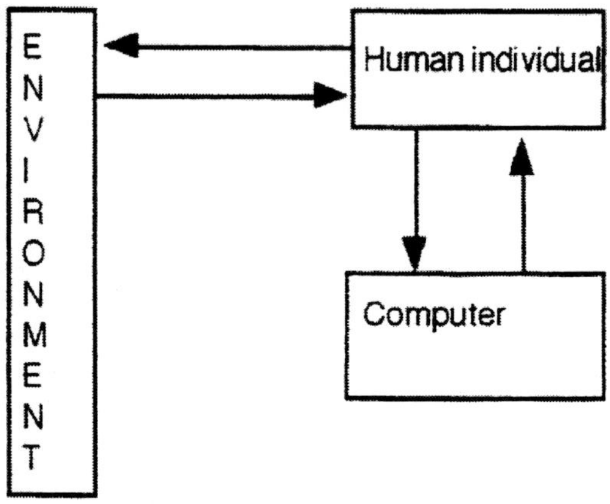

Figure 1. An elementary system (from Rasegård, 1998).

A human individual in this example is working with the help of a computer. The human individual takes in information from the environment. This information he refines in the computer from which he receives feedback of arranged information which he thereafter delivers to the environment. As can be seen, the definition has not mentioned the influence of environment and feedback, both of which are components of essential importance for a system which involves human activities. However, the definitions of systems are not complete without mention of the division into open and closed systems. A citation that brings light to this division is given in the following: "Let us look a little more closely at the individuality of an organism. It is clearly, not quite the same as the individuality of (say) a diamond or a piece of hard metal. These pieces of solid matter are crystals. They are systems of oscillating atoms, atoms which in the main are neither leaving not joining the system within a fairly long period of time: they are closed systems – closed with respect to the material particles of which they consist. (Though they are open with respect to energy flow.) By contrast, organisms are open systems – like flames. They exchange material particles (and also, of course, energy) with the environment: they have metabolism. Nevertheless, they are identifiable individuals. They are, as indicated by Locke, identifiable even during growth: they are identifiable dynamic processes: or perhaps better, material systems undergoing exchange of matter. When we speak of an organism, we often forget this, because during a sufficiently short time span an organism is approximately closed, almost like a crystal". (Popper and Eccles, 1981, pp. 113-114).

The main difference between open and closed systems is shown in Figure 2.

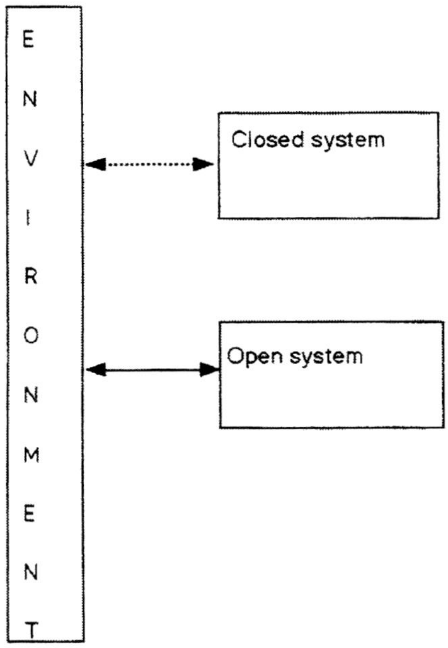

Figure 2. The main difference between a closed and an open system (from Rasegård, 1998).

The figure shows that the closed system has nil or very weak exchange of influences vis-à-vis the environment, which the dotted double arrow indicates while the solid double arrow indicates continuing influences vis-à-vis the environment. This makes the distinction between closed and open systems. However, it is also possible to make another division into two main categories: Natural systems, for example, the solar system; and man-made systems, for example, the managing system of an organization.

The wide meaning of the system concept is indicated by the following citation: " The term system has a number of meanings. There are systems of numbers and of equations, systems of value and of thought, systems of law, solar systems, organic systems, management systems, command and control systems, electronic systems, even the Union Pacific railroad system. The meanings of 'system' are often confused. The most general, however, is: A system is a set of interacting units with relationships among them. The word set implies that the units have some common properties. These common properties are essential if the units have to interact or have relationships. The state of each unit is constrained by, conditioned by, or dependent on the state of other units. The units are coupled. Moreover, there is at least one measure of the sum of its units which is larger than the sum of that measure of its units." (Miller, 1978, p. 16). This citation indicates the *synergetic* effect of a system where the sum of the measure of the whole system is larger than the sum of each of its parts taken together.

1.3. Some Further Comments on the System Concept

"A system is an object or processes – or collection of objects or processes – that we choose to study. The environment comprises those processes that we expect to have an effect on the system. The inputs to the system are the effects by the environment on the system. The outputs of the system are the measurable responses of the system. The states of the system relate the inputs and outputs. The state-transition function describes how the states change over time in response to inputs. The output function describes how the states determine the output." (Arbib, 1977, p. 176). This statement describes a system as an object – or collection of processes that we choose to study. This means that a boundary of a system depends on what an observer desires to study. And the components within this boundary are those which have a special interest to the observer. The system's environment is also influencing the system through the inputs, and the system's refining of inputs in measurable outputs (goods and services)

Even if the concept of system in the foregoing has been seen from different aspects , it seems necessary to point out certain features which might not have been pointed out in a satisfying way earlier. A system must be equipped with *requisite variety* or, in other words, it must be capable of dealing with different states influenced by the environment. For example, a system on the level of organization must be able to cope with different influencing factors from the outside world, e.g., competitors' , suppliers' and customers' moves and measures on the commercial market. This means that the human individuals in the system must have necessary competence to handle various situations and be equipped with artifacts which respond to demands of a working rationale. Further, a system must be recursive . That means, for example, in a system on the level of organization with subsidiaries, that the superior system structure is found in the organization's different parts, for example subsidiaries, departments and sections. In other words, it would in principle be something like a Russian doll or a Chinese box.

As mentioned above, a condition for a viable system is that it is recursive. But ought the knowledge about the environmental conditions for the system's survival and development be the same on the different levels of recursion ? No, it is quite natural that the knowledge about the organizational system's conditions and its different environmental connections will be more profound on the higher organizational levels. However, this does not mean that the professional competence of the human individuals needs to be lower on lower levels. What is said here can in some way be read out of the following: "On the lowest epistemological level, referred to as level 0, a system is defined by a set of variables, a set of potential states (values recognized for each variable), and some operational way of describing the meaning of the variables and their states in terms of the associated real world attributes and their manifestations. The set of variables is always partitioned into two subsets, referred to as basic and supporting variables. Aggregate states of all supporting variables form a support set within which changes in the states' basic variables occur. The most frequent examples of supporting variables are those representing time, space, or various populations of individuals of the same kind. The term source systems is usually used for systems defined at this level to indicate that such systems are, at least, potentially, sources of empirical data.

Systems on different higher epistemological levels are distinguished from each other by the level of knowledge regarding the variables of the associated source system. A higher-level system entails all knowledge of the corresponding system on any lower levels and, at the same time, contains some additional knowledge that is not available on the lower levels. Hence, the source system is included in all of the higher-level systems." (Klir and Folger, 1988, pp. 194-195).

At last there can be reason to refer to what is the innermost steering factor of a system which involves both expectations and disappointment caused by human individuals: " So the idea of the genetic program is that living organisms not only have a matter of which they are made up, but also something that organizes them; the genetic program. It is of course unclear what the genetic program actually is, but anyway this is what is supposed to give the organisms their meaning, significance and purpose. It is very much an Aristotelian kind of concept, a teleological concept which has been smuggled back into biology. It is often decked out in the terminology of Information Theory, and talked about in terms of genetic information. Its dualistic nature comes out in the well-known hardware/software analogy, based on computers. The proteins and so on in the cells are the hardware; the program is the software which controls them."

(Rupert Sheldrake in *The Search for Meaning*, edited by Pylkänen, 1989, p. 95).

Chapter 2

CERTAIN ASPECTS OF MAN´S RELATIONS TO DIFFERENT SYSTEMS

2.1. FROM MICRO TO MACRO

Man is living encapsulated in systems . To give detailed accounts of all these systems is not possible. Therefore the description in this report is limited only to give an overview of the existence of certain natural systems, and a more detailed description of certain types of man-made systems, which the author considers as being the most important. However, in order to show a more compressed picture of the most significant features, some basic elements building up systems from micro to macro, the following figure by Professor John D.Barrow can be used as an illustration.

Here follows Professor Barrow´s comments on the figure: "In the figure are shown the most significant structures in the known Universe. They range from the nucleus of the hydrogen atom, larger atoms and molecules, to people, trees and mountains to asteroids, planets and stars, before extending to the largest astronomical structures we see, galaxies and clusters of galaxies, and then to the entire visible Universe. There is a 'magic' box that can be drawn on this diagram. It contains the range of structures which exhibit properties characteristic of organized and complex systems. That is, their structure is not simply dictated by the balance between two opposing forces of Nature. Their essential nature is a consequence not of their size alone but of the manner in which their internal constituents are organized. You will notice that human beings sit four-square inside the magic box. This it not surprising. Our brains are the most complicated objects that we have so far encountered in the Universe. We are far from simple. Indeed, were our brains significantly simpler, we would be too simple to know it" (Barrow, 1991, pp. 136-137).

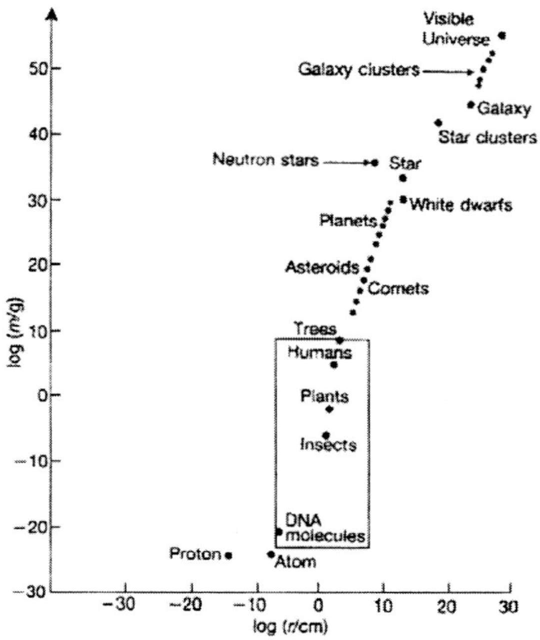

A hierarchy of complexity. Legend: The spectrum of sizes and masses of the principal structures known to exist in the Universe, with the region in which complex organized structures form highlighted by the box. Within this region the phenomena observed are primarily the result of the complex interactions between very large numbers of mutually interacting constituents. They are the results of particular forms of organization between those components, rather than the unusual behavior of individual forces of Nature alone. (from Theories of Everything, p. 137).

Figure 3. A hierarchy of complexity (from Barrow, 1991).

2.2. Man and Different Systems

The figure below (Fig.4) can be seen as an attempt to illustrate man´s relation to different types of systems in a simple diagram. As we can see, overlapping cosmic systems ultimately have strong influence on other subordinated systems. As an example, can be mentioned the eruptions on the Sun, which affect the climate on the Earth. However, a prerequisite for all systems are elementary systems consisting of molecules, atoms and sub-particles of atoms. These elementary systems are influencing both the overlapping cosmic systems and the Gaia systems. The latter denotation is taken from the British researcher James Lovelock's book "Gaia" which is the Grecian name of the Goddess of the Earth. The denotation Gaia systems is a comprehensive term for the whole web of natural systems on our little planet in the Universe. Because the Earth is a living planet, it is populated by living systems of different kinds and degrees of development. Further, we can observe that living systems are divided into three classes. This dividing has been influenced by von Bertalanffy's table on pp. 28-29 in "General Systems Theory." From this table, it seems possible to, in a simplified way, derive three main types of living systems, namely, automatic living systems (lower organisms, plants etc), instinctive intelligent living systems (animals) and higher intelligent living systems (man). Automatic living systems have reciprocal connections to higher intelligent living systems, represented by man.

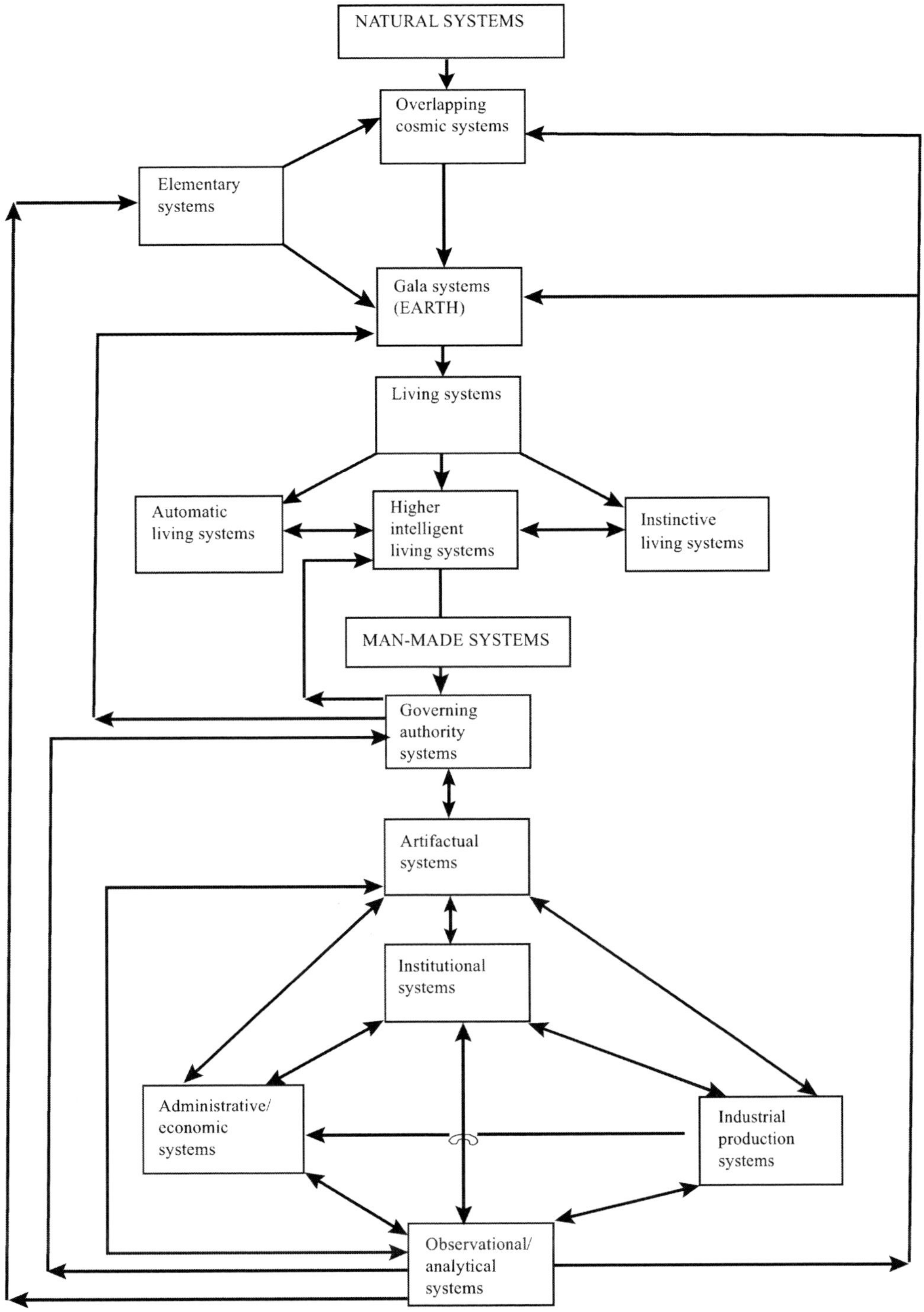

Figure 4. Man in a web of systems (from Rasegård, 2002).

Automatic living systems influence man's conditions of life, for example, food supplies in the form of crops of different plants, not the least grain, and feeding stuffs for domesticated animals. Man in turn influences the systems, for example, by agricultural measures in order to improve the crops. However, man is also responsible for danger in the form of over-exploitation and using of poisoned chemicals for combating weeds and using gene manipulation out of control. When we come to instinctive intelligent living systems, it can be noted these systems' importance for man, for example, production of foods based on animal products. However, even here it is a danger of gene manipulation out of control in order to increase the production of animal-based foods. Another danger in this connection is the use of unsuitable food stuff for domesticated animals, which can even infect human organisms, for example, BSE-disease. Man also has an ultimate responsibility for shaping the balance between different kinds of wild animals, predators as well as prey. Even if what has been said here is widely known and perhaps can be denoted as banalities, it seems necessary to use words to frame yhe importance of living systems and man's responsibility for not disturbing the balance of these systems as a consequence of his greed.

2.3. Man-Made Systems

Now we leave the systems which have been denoted as natural and enter the systems made by man. In this section of man-made systems, governing authority systems are superior to all other systems. Under this term all systems of overlapping character are included. As examples of such systems can be mentioned governmental systems for ruling a country consisting of parliament, government and subordinated organs, of course, even supranational organizations influencing a country are here included. These governing authority systems also influence Gaia systems by decisions made concerning environmental aspects, for example, dam projects for power and irrigation. However, decisions made by the governing authority systems also influence the living of higher intelligent systems represented by man through issued laws and ordinances. These laws and ordinances also influence man's relations to automatic living systems as well as instinctive living systems, for example, prohibition laws of using poison for combating weeds and certain insects and painful medical experiments on animals, for example, apes. Of course, through governing authority systems, issued laws and ordinances also regulate the mutual relations between the human individuals as well as with the society in which they live. However, the individuals who shall carry out their duties within the mentioned systems can not do their work without using artifacts, e.g., devices of different types, sometimes coupled together and forming systems. However, a society of mankind also contains institutional systems of different kinds, for example, educational, technical, cultural etc. These institutional systems influence the society's activities in different ways. As we can see both administrative/economic systems as well as industry production systems are influenced by institutional systems. But these types of systems also give impulses back to the institutional systems, for example, in the form of demands for improved or new education for administrators and technicians. Administrative/economic systems have reciprocal connections to industry production systems. The former systems are responsible for economic and administrative managing of the latter systems while these systems provide economic /administrative systems with information about production. Thus, administrative/economic systems register reports from industry production systems and in turn deliver work-up reports to the latter systems.

Observational/analytical systems, are systems that give impulses of improvements and progress of human life. Within these systems are systems of scientific research and also systems for improving activities within the administrative/economic systems included, as well as within industry production systems and institutional systems. As we can see, these systems besides the mentioned types of systems have connections to other systems. To begin, we can see that observational/analytical systems influence governing authority systems through analyses of these systems' functions, which can result in proposals of improvements. We can also see (the loop in the left of the figure) that observational/analytical systems can influence the elementary systems. How is that possible? The answer is that the observational/analytical systems can provide the scientific world with more effective organizations of research which means that the research in elementary systems can be more sharpened through liberating researchers from a deal of bureaucratic complications. In the loop in the right of the figure, we can see that observational/analytical systems are connected to Gaia systems as well as the overlapping cosmic systems. These connections indicate, as have been said earlier, the communities of science research works within observational/analytical systems. These research works in turn influence subordinated systems. However, we must not forget the importance of artifactual systems. As we can see, these systems have reciprocal connections with all man-made systems. Thus, impulses for improvements come from all of these systems and are realized in the form of improved and new artifacts and systems mainly carried out by factories and workshops within industry production systems.

Perhaps, it can be of interest to look at the following citation from the British biologist James Lovelock: "What is remarkable about man is not the size of his brain, no greater than that of a dolphin, nor his loose incomplete development as a social animal, nor even the faculty of speech or his ability to use tools. Man is remarkable because by the combination of all these things he has created an entirely new entity. When socially organized and equipped with a technology even as rudimentary as that of a stone age tribal group, man has the novel capacity to collect, store and process information, and then use it to manipulate the environment in a purposeful and anticipatory fashion."(Lovelock, 1979, p. 132). Maybe, we can continue the citation from the same author: "From a Gaian viewpoint, all attempts to rationalize a subjugated biosphere with man in charge are as doomed to failure as the similar concept of benevolent colonialism. They all assume that man is the possessor of this planet, if not the owner, then the tenant. The allegory of Orwell's 'Animal Farm' takes a deeper significance when we realize that all human societies in one way or another regard the world as their farm. The Gaia hypothesis implies that the stable state of our planet includes man as a part of, or a partner in a very democratic entity."(Lovelock, 1979, p. 145).To ignore what is said here will mean the suicide of man.

Maybe, it can be appropriate to end this section with the following words from Paul Davies, the well-known British Professor of Astronomy: "I have been at pain to argue the organizational principles needed to supplement the laws of physics that are likely to be forthcoming as a result of new approaches to research and new ways of looking at complexity in nature. I believe that science is in principle able to explain the existence of complexity and organization at all levels including human consciousness, though only by embracing the "higher level" laws. Such a belief might be regarded as denying a god, or a purpose in this wonderful creative universe we inhabit. I do not see it that way. The very fact that the universe is creative, and that the laws have permitted complex structures to emerge and develop to the point of consciousness - in other words, that the universe has organized its own self-awareness – is for me the powerful evidence that there is "something going on" behind it all. The impression of design is overwhelming. Science may explain all the processes whereby the universe evolves its own destiny, but that still leaves room for there to be a meaning behind existence."(Davies, 1987, p. 203).

Chapter 3

ARTIFACTUAL SYSTEMS

However, an open system, or it can also be described as a living system, is often depending on devices, so-called artifacts, in order to be able to operate in a satisfying way. A definition of the concept of artifacts is given below according to the following: "An artifact is an inclusion in some system, made by animals and man. Spider webs, bird nests, beaver dams, houses, books, machines, music, paintings and language are artifacts. They may or may not be prostheses – inventions which carry out some critical process essential to a living system. An artificial pacemaker for a human heart is an example of an artifact which can replace a pathological process with a healthy one. Insulin and thyroxine are replacement drugs which are human artifacts. Chemical, mechanical, or electronic artifacts have been constructed which carry out some functions of all levels of living systems." (Miller, 1978, p. 33).

Figure 5 is meant to illustrate some components´ connections in an artifactual system related to the economic management of an organization. As can be seen, this artifactual system is manipulated by human individuals. Of course, the system in itself does not have any value in the management, but only when it is operated by human individuals. When not operated it can be denoted as a resting system which is aroused by human individuals. These individuals who have contacts with the environment, as an end-product receive reports which can influence the environmental contacts. Further explanations of the concept of artifacts are given in the citation below.

> "Music is a special sort of human artifact. So are the other arts and cognitive systems which people share. So is language. Whether it be a natural language or the machine language of some computer system. It is essential to information processing often stored in human brains and expressed only by human lips. Language can also be recorded on non living artifacts, like stones, books, and magnetic tapes. It is not of itself a concrete system. It changes only when man changes it. As long as it is used, it is in flux, because it must remain compatible with the ever-changing living systems that use it. But the change emanates from the users, and without their impact the language is inert."
>
> (Miller, 1978, pp. 33-34).

Some more examples of artifacts related to the citations above are given in the following figure 6.

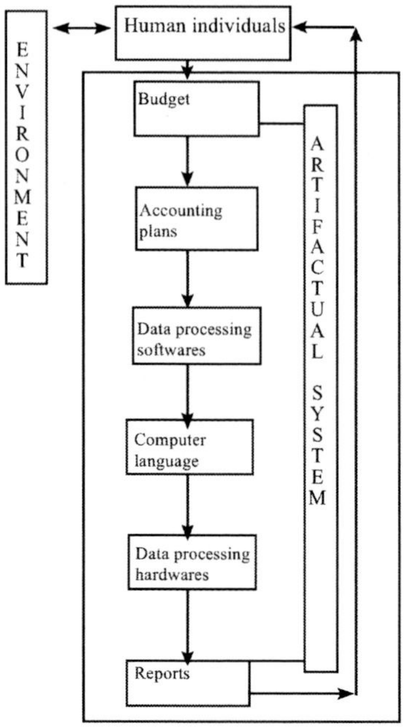

Figure 5. An example of an artifactual system (from Rasegård, 1998).

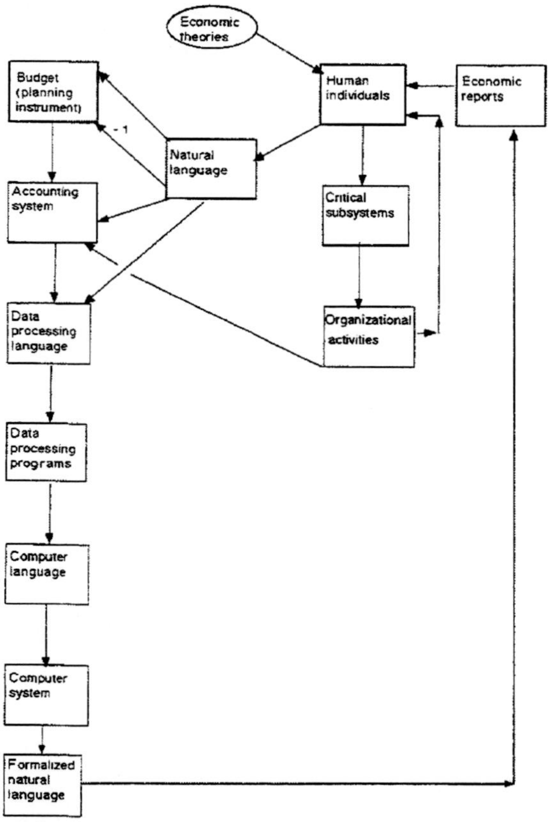

Figure 6. Theories and language in an economic system (from Rasegård, 1998).

The economic theories influence the human actions which use the natural language transformed into written language in the work of designing the budget and the accounting system, which is registering the organizational activities and is the base for further data processing by the computer system. This means that the accounting system influences the data processing language. The data processing programs are written in a modified natural language in order to be able to communicate with the computer system and its computer language. This computer system produces reports written in a formalized written natural language. These reports are delivered to the human individuals and can lead to a revision of the budget, which is indicated by the arrow – 1. (The budget, namely, is the overriding governing economic document). In this figure it is also presumed that the organizational tasks are divided according to the principles of the 20 critical subsystems described in the next section.

> "Scientists sometimes neglect to distinguish between living systems and their artifacts. Because artifacts are the products of living systems, they often mirror aspects of their producers and thus have systems characteristics of their own. Termites nests, pots and jewelry of primitive tribes, and modern buildings are all concrete systems which can be studied for themselves alone as well to understand the living systems that produced them. But they themselves are not living systems. Systems theory can also be applied to the history, dynamics over time, rules of change, and other aspects of languages or music, if they are viewed as abstracted systems independent of the living system that produced and used them. This may be desirable because their producers or users may be long dead or unavailable for study".
>
> (Miller, 1978, p. 34)

Thus, the artifactual system exemplified in figure 6 can be said, in some way, to be a mirror of the organization´s structure and its activities. The artifactual system seen as a separate unit thus can be regarded as a concrete system. But in a historical perspective it can also be viewed as an abstracted system of historians of, for example, economic history. These researchers can, through the system, obtain insight of how the economic results of these activities were organized, the economic results of these activities and the influencing factors of the outcomes.

Chapter 4

AN OBSERVATIONAL/ANALYTICAL SYSTEM

4.1. AN OUTLINE OF INVESTIGATIONS

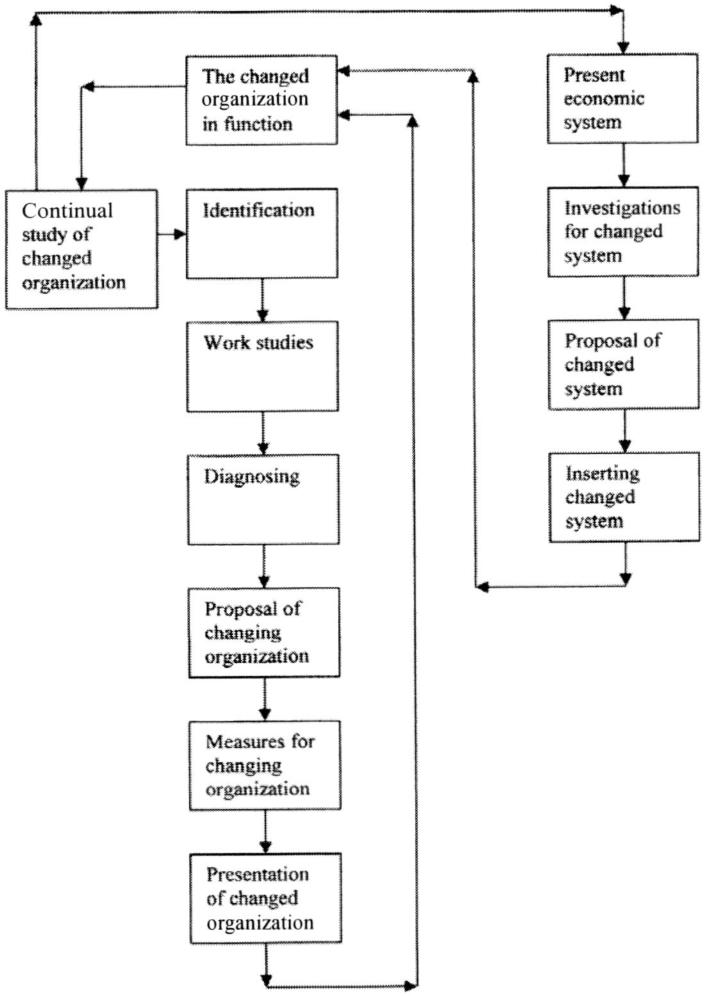

Figure 7. Main steps of organizational investigations.

The figure above shows the main steps of an organizational investigation which begins with the identification of the critical subsystems′ positions in the organization. A presentation of these subsystems is done in the next section. When this is done it is time for a study of each subsystem′s works considering duties and cooperation with other subsystems. This study is done with the help of job descriptions completed with interviews and personnel observations. After this study it is time for diagnostic discoveries of actual malfunctions. This can be done using the questionnaire which is shown later on in this report. Since the diagnosis is done it is time to work out the proposal for a changed organization. At last the proposal for a changed organization has to be presented to the decision makers for final decision.

However, it is not enough to study a structure of an organization. A very important part of an organizational study is to have a look at the economic system. Here it can be a question of mapping the present system for forming a base for judgment if the system is up-to-date, if not, it is necessary to start the investigation of possibilities to substitute the present system with some other. In this case a system built on the principles of zero-based budgeting can come into focus. The principles of such a system will be clarified later on in this report.

However, even if an improved or new structure of an organization and its economic system has come into being it is very important that responsible managers are aware of the development of the organizational systems as well as economics. That means that a system must be under continual observation. In this connection a Swedish professor of economics can be cited when his course comes to an end: “ Now we have finished this course, but anyone who thinks he has finished he is finished”. That means that the learning function, the will to learn something new, always shall be present.

4.2. Emotional Intelligence Influencing Team Work

In connection with the foregoing section which mentioned investigation work it can be of interest to point the concept of emotional intelligence described by Dr. Daniel Goleman in his book with the same name. He states that besides our conventional IQ we also have to take into consideration the influence of emotional intelligence. This form of intelligence then is seated in the amygdala (from the Greek word for “almond”:

> “It is an almond-shaped cluster of interconnected structures perched above the brainstem, near the bottom of the limbic ring. There are two amygdalas, one on each side of the brain, nestled toward the side of the head. The human amygdala is relatively large compared to that in any of our closest evolutionary cousins, the primates.
>
> The hippocampus and the amygdala were the two key parts of the primitive “nose brain” that, in evolution, gave rise to the cortex and then to the neocortex. To this day these limbic structures do much or most of the brain′s learning and remembering: the amygdala is the specialist for emotional matters. If the amygdala is severed from the rest of the brain, the result is a striking inability to gauge the emotional significance of events; this condition is sometimes called “affective blindness”. In the following figure an attempt is made to show how certain factors connected to IQ and emotional intelligence influence a team work of investigations and scientific research. Whenever people come together to collaborate whether it can be in an executive planning meeting or as a team working toward a shared product, there is a very real sense in which they have a group IQ, the sum total of the talents and skills of all those involved. And how well they accomplish their task will be determined by how

high that IQ is. The single most important element in group intelligence, it turns out, is not the average IQ in an academic sense, but rather in terms of emotional intelligence. The key to high group IQ is social harmony. It is this ability to harmonize that, all other things being equal, will make one group especially talented , productive and successful, and another – with members whose talent and skill are equal in other regards – do poorly. The idea that that is a group intelligence at all comes from Robert Sternberg, the Yale-psychologist, and Wendy Williams, a graduate student, who are seeking to understand why some groups are far more effective than others" (Goleman, 1997, p. 183).

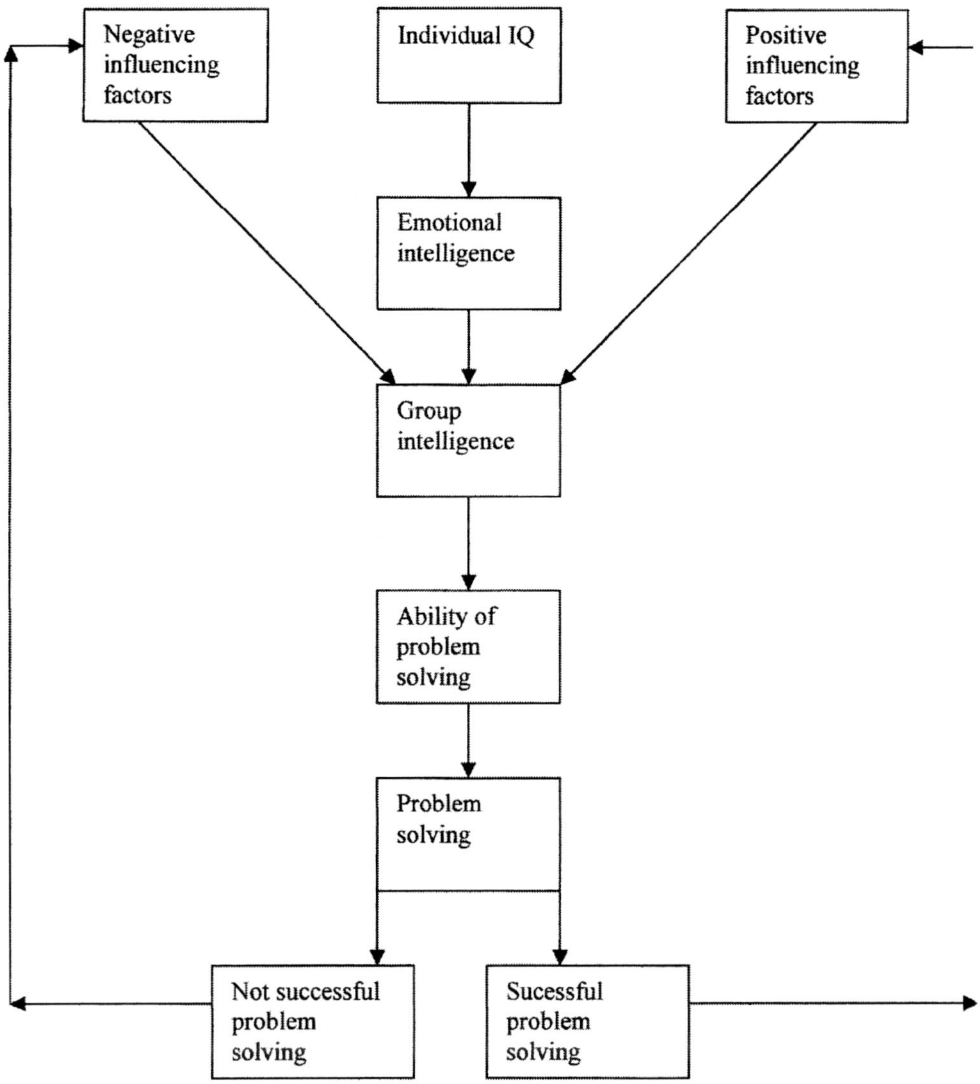

Figure 8. Emotional intelligence influencing team work.

Taken the idea of emotional intelligence for granted, we can have a look at the following figure. As we can see the figure the individual IQ is presumed with emotional intelligence. This type of intelligence as well as the conventional IQ form the group intelligence. The conventional IQ can vary between the different individuals in the group as the emotional intelligence. However, the emotional intelligence is important if a research or an investigation team wants to be effective in its work. The same is true of course in all sorts of team work. Presume that a person of a high IQ who not so well-developed emotional intelligence can, for example, pester the whole group with his sarcasms about some person who he considers to have an IQ matching his. This will have a negative influence on the group intelligence. On the other hand a person with not so high an IQ but with a developed emotional intelligence can serve as an inspiration for the other members of the group and so influence the group intelligence in a positive direction. Therefore it is a self-evident truth that the ability to problem solve can be damaged by persons who have a lesser developed emotional intelligence. The problem solving leads to not so good of a result which is only reinforcing the bad influences on team work if the disturbing source is not taken away. On the contrary a group who has high group intelligence including emotional intelligence is presumed to obtain good solutions of problems which in turn influences the team work of research in a positive way. Thus, the figure below gives a hint of certain problems coupled to the term "group intelligence." It can be added that the author of this report has been acquainted with the said problem when taking part in investigation and research teams concerning management and organization matters in societal organizations. Nevertheless, it can be remarked that it is not unusual that scientific researchers who have a remarkably high IQ can prefer to do their research work alone as some kind of "lonely wolves."

4.3. Presentation of an Observational/Analytical System

This example of a system is a system composed of human individuals or, in other words, a living system in contrast to the former examples. The system is based on the theory of 20 critical subsystems created by the American Professor James Grier Miller). The functions of these subsystems are shown in figure 9.

The symbol of each of these subsystems are shown in the next figure (figure 10). These symbols have been worked out by the Swedish Professor Kjell Samuelson of Stockholm University and Royal Institute of Technology in cooperation with Professor James Grier Miller, author of *Living Systems*.

In the following example, an attempt is made to apply the critical subsystems to a community of science. The subsystems' profiles and main functions connected to the scientific community are exemplified according to the following. The example is taken from an earlier published paper by the author of this report.

Functions in an Organization

Subsystems Which Process Both Matter-Energy and Information

1. *Reproducer*, contains the society's educational system with personnel who educate people in order to obtain such a level of knowledge that they have the possibilities of entering institutions for scientific education.

THE 20 CRITICAL SUBSYSTEMS OF A LIVING SYSTEM	
SUBSYSTEMS WHICH PROCESS BOTH MATTER-ENERGY AND INFORMATION **1.** *Reproducer*, the subsystem which carries out the instructions in the genetic information or charter of a system and mobilizes matter, energy, and information to produce one or more similar systems.	
2. *Boundary*, the subsystem at the perimeter of a system that holds together the compnents which make up the system, protects them from environmental stresses, and excludes or permits entry to various sorts of matter-energy and information.	
SUBSYSTEMS WHICH PROCESS MATTER-ENERGY	**SUBSYSTEMS WHICH PROCESS INFORMATION**
3. *Ingestor*, the subsystem which brings matter-energy across the system boundary from the envionment.	**11.** *Input transducer*, the sensory subsystem which brings markers bearing information into the system,changing them to other matter-energy forms suitable for transmission within it.
	12. *Internal transducer*, the sensory subsystem which receives, from subsystems or components within the system, markers bearing information about significant alterations in those subsystems or components, changing them to other matter-energy forms of a sort which can be transmitted within it.
4. *Distributor*, the subsystem which carries inputs from outside the system or outputs from its subsystems around the system to each component.	**13.** *Channel and net*, the subsystem composed of a single route in physical space, or multiple interconnected routes, over which markers bearing information are transmitted to all parts of the system.
	14. *Timer*, the subsystem which transmits to the decider information about time-related states of the environment or of compunents of the system. This information signals the decider of the system or deciders of sybsystems to start, stop, alter the rate, or advance or delay the phase of one or more of the system's processes, thus coordinating them in time.
5. *Converter*, the subsystem which changes certain inputs to the system into forms more useful for the special processes of that particular system.	**15.** *Decoder*, the subsystem which alters the code of information input to it through the input transducer or internal transducer into a "private" code that can be used internally by the system.
6. *Producer*, the subsystem which forms stable associations that endure for significant periods among matter-energy inputs to the system or outputs from its converter, the materials synthesized being for growth, damage repair, or replacement of components of the system, or for providing energy for moving or constituting the system's outputs of products of information markers to its suprasystem.	**16.** *Associator*, the subsystem which carries out the first stage of the learning process, forming enduring associations among items of information in the system.
7. *Matter-energy storage*, the subsystem which places matter or energy at some location in the system, retains it over time, and retrieves it.	**17.** *Memory*, the subsystem which carries out the second stage of the learning process, storing information in the system for different periods of time, and then retrieving it.
	18. *Decider*, the executive subsystem which receives information inputs from all other subsystems and transmits them to information outputs for guidance, coordination, and control of the system.
	19. *Encoder*, the subsystem which alters the code of information input to it from other information processing subsystems, from a "private" code used internally by the system into a "public" code which can be interpreted by other systems in its environment.
8. *Extruder*, the subsystem which transmits matter-energy out of the system in the form of products or waste. **9.** *Motor*, the subsystem which moves the system of parts of it in relation to part or all of its environment or moves components of its environment in relation to each other.	**20.** *Output transducer*, the subsystem which puts out markers bearing information from the system, changing markers within the system into other matter-energy forms which can be transmitted over channels in the system's environment.
10. *Supporter*, the subsystem which maintains the proper spatial relationships among components of the system, so that they can interact without weighing each other down or crowding each other.	

Figure 9. The 20 critical subsystems of a living system (from Behavioral Science, vol. 36, 1991).

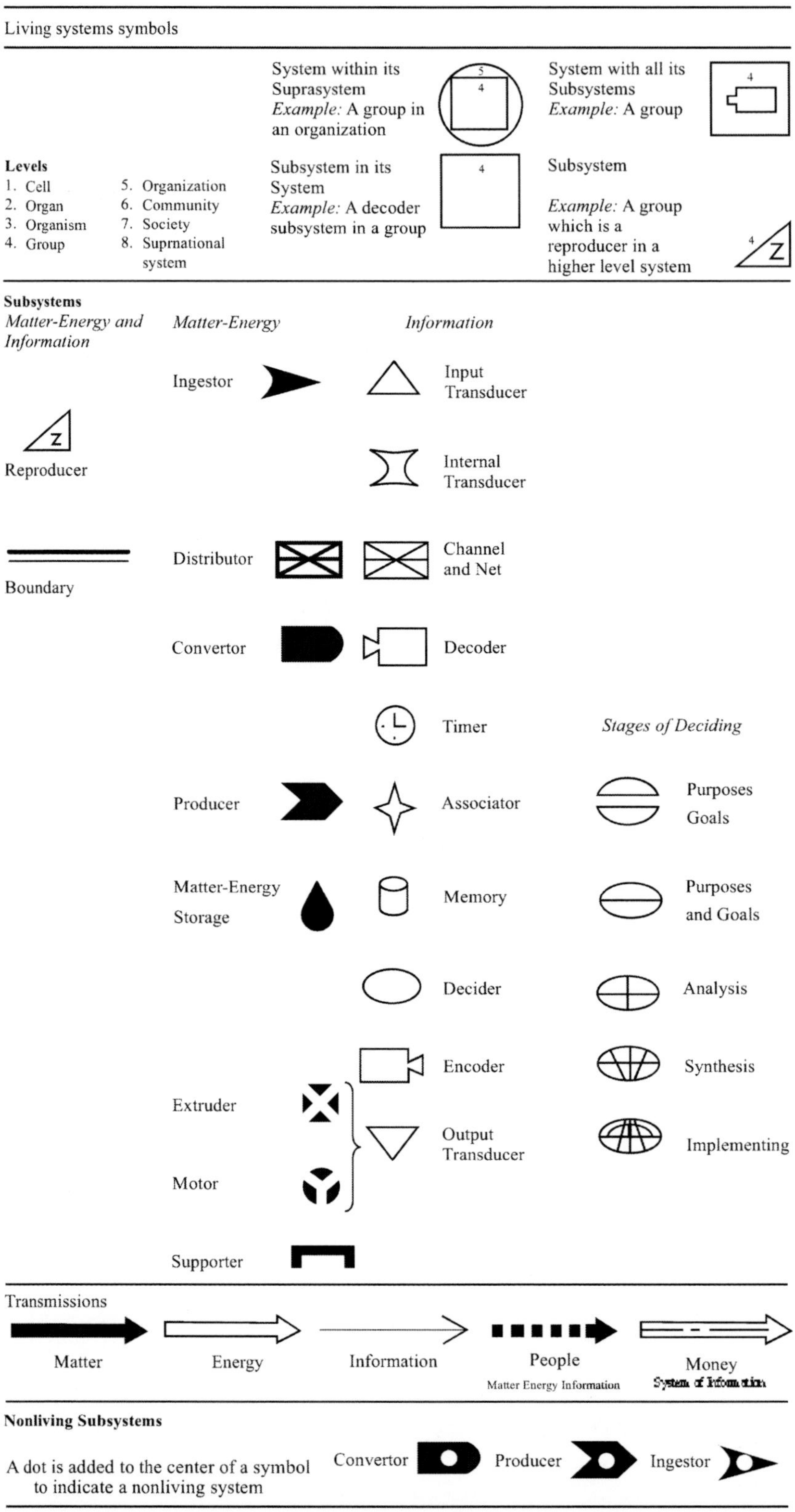

Figure 10. Living systems symbols (from Behavioral Science, vol. 36, 1991).

2. *Boundary*, can be divided into physical boundary and information boundary.
   Physical boundary, buildings and fences belonging to scientific institutions (universities and other institutions of scientific research).
   Information boundary, authorities and other authorized organs which make investigations of applicants' merits for entering the community of science as researchers as well as other necessary personnel involved in the activities of the community.

Subsystems Which Process Matter-Energy

3. *Ingestor,* purchase departments at the scientific institutions which are responsible for purchase of materials, artifacts and other necessities for research activities, personnel departments who are employing personnel to work within the community of science.
4. *Distributor,* transport means which carry material, artifacts and other necessities from outside the system or between the units within the system, for example, transport of laboratory tests from one institution to another.
5. *Converter,* instrument shops at certain scientific institutions which adjust and refine purchased instruments in order to make them more suitable for special scientific purposes, the catering activities which convert food stuffs making them suitable for making meals at universities' and other scientific institutions' bars and restaurants.
6. *Producer,* laboratories at institutions of natural sciences which are working out scientific tests of hypotheses to be verified, and scientific controls of certain stuffs as foods, chemicals etc., workshops for making instruments and other devices for use in the scientific work, bars and restaurants created for personnel at universities and other scientific institutions.
7. *Matter-energy storage,* storing of artifacts, materials and other necessities for use in the scientific work.
8. *Motor,* the transport system which carries personnel into and out of the scientific working-places.
9. *Extruder,* cleaning and destruction units within universities and other scientific institutions.
10. *Supporter,* the building construction of universities and other scientific institutions.

Subsystems Which Process Information

11. *Input transducer*, researchers and scientific journalists reporting from scientific seminars and conferences, scientific attachés at foreign embassies, reviewers of scientific books and papers.
12. *Internal transducer,* units of economic accounting, personnel planning and internal information, personnel responsible for computed data bases.
13. *Channel and net,* telephone interchanges at scientific institutions and computed data-net with operators, scientific conferences and informal contacts between scientists, manual messenger services.

14. *Timer,* personnel who set timetables for scientific experiments and research projects and their artifacts in the form of clocks and signal devices of time.
15. *Decoder,* judicial personnel who interpret laws and governmental ordinances in such a form that it can be available for scientists, for example, ethical and ecological legislation, rules of applications and administration of research project, judicial agreements between universities and other scientific institutions visa `a vis the outer world, for example, research projects financed by industrial companies, translators of scientific books and articles from foreign languages.
16. *Associator,* scientific councils of different branches of science which investigate projects of scientific researchers judging their value for receiving grants or other support. These councils serve also as advisers to the scientific and political decision makers. This subsystem is also the center for theoretical scientific research and creativity when it comes to forming hypotheses and verifying these - in other words, such scientific activities which do not require processing matter-energy. This last part, namely, is carried out by the subsystem producer.
17. *Memory,* scientific libraries, archives, diaries and registers (both manual and computed), and computed data bases of various kinds.
18. *Decider,* boards of universities and other scientific institutions, government and its organ to which certain decisions are delegated.
19. *Encoder,* personnel who prepare scientific reports for scientific journals, international conferences, political decision makers and for mass-media and popular scientific use out of the community of science.
20. *Output transducer,* editors of scientific journals and other personnel who edit the information material prepared by encoders and who have the responsibility of publishing and contacts with mass media representatives and other representatives of the environment.

The connections between the different subsystems are shown in the next section.

4.4. Chart of Positions Showing the Cooperation of Subsystems

A. Information Flow

Necessary information is taken into the community by *11. Input transducer.* This information is prepared and arranged by *12. Internal transducer* which among others scrutinizes the information from a pure scientific viewpoint and is completed with information from the different units within the community stored in *17. Memory,* which in this case is divided into two parts , one for practical matters connected to the managing of science (17:1) and one for pure scientific research(17:2). The former is a data base of, for example, administrative rules, judicial documents etc. while the latter, as has been said, is for pure scientific information. The prepared and arranged information is relayed to *15. Decoder* and *16. Associator* and is in certain cases directly relayed to the units of natural research within the matter-energy section. In addition certain information concerning, for example, applications employing researchers and other personnel as well as need of supplies is conveyed to *3. Ingestor. 15.Decoder* is responsible for making the information useful for internal use and makes clear how the information from the outer world should be judged. Among other things it also makes information suitable for the computed data processing where it is processed and stored in *17. Memory* for further access. The information interpreted and arranged by *15. Decoder* is also relayed to *16. Associator* where it will be put together for use by *18. Decider* . Besides that, *15. Decoder* is relaying information to the units of natural sciences within the matter-energy section. From these units in certain cases are received questions concerning further information, for example, clarifications of judicial matters. *16. Associator* arranges and makes syntheses of received information which is forming the basis for decisions by *18. Decider*. This information then has been received from *12. Internal transducer* (arranged information from the internal environment), *15. Decoder* (arranged information from the external environment) and *17. Memory* (17:1-17:2). This processed information in suitable parts is relayed to *17. Memory* as feedback. Thus, from *17. Memory,* information taken in an earlier atage now will be completed with further information.

18. Decider consists of two types of decision makers marked 18:1 and 18:2.The former is the decider connected to the community of science who makes decisions about, among others, allocations of grants and funding sources and also makes decisions about personnel and other practical matters within the community. The latter is the level of political decisions where the research policy in the whole is determined. At this level it sometimes can be a question of giving priority to a certain branch of research in relation to others. The foundations for these decisions are received from committees connected to the associator function. These foundations can in certain cases be passed back to the respective committee for revision. This is marked by the feedback loop from *18:2 Decider* and further to *16. Associator*. From *18:1 Decider* is also drawn an information loop to *12 .Internal transducer* for forwarding the measures to be taken within the community. *19. Encoder* prepares and edits scientific materials and other reports to the outer world. In this connection it is taking the scientific information stored in *17:2 Memory* while information about administrative matters are received from personnel connected to the decider function. *20. Output transducer* is responsible for bringing out the scientific material and reports arranged by the encoder function to other communities of science, mass media, authorities and citizens. It can be noted that *19. Encoder* and *20. Output transducer* sometimes can be coupled together into one function. The same can be said of *12. Internal transducer* and *15. Decoder.*

13.Channel and net connects all of the subsystems and is in the diagram marked by thin lines between the subsystems.

B. Matter-Energy Flow

3.Ingestor takes into the community material, artifacts and other necessities as well as personnel needed within the system. This subsystem has, earlier, from *12. Internal transducer* received information about the actual needs of personnel, artifacts and other necessities. *7. Matter-energy storage* stores artifacts and other necessities for use in the community of science. The registration of the articles in store is made in the *17:1. Memory*.

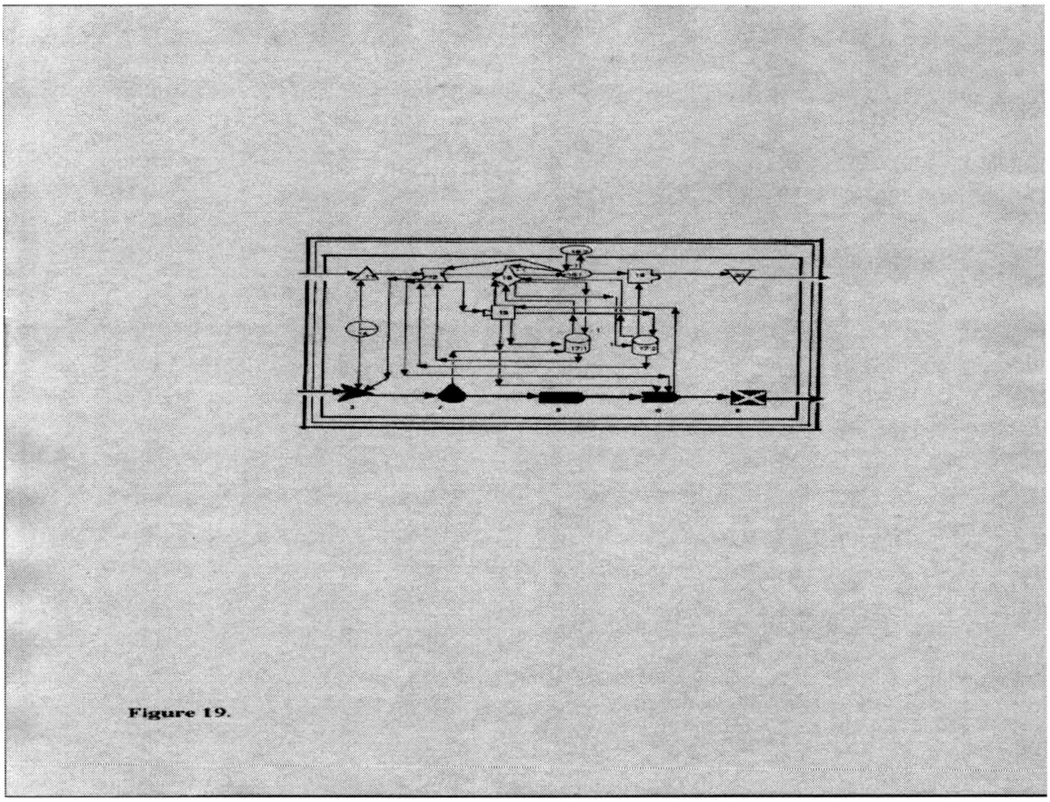

Figure 11. An outline of the critical subsystems′ positions in a community of science (from Rasegård, 1997).

5. Converter adjusts materials and other necessities so it can be used for special purposes. *6. Producer* is the system which is working with scientific experimental research connected to natural sciences within the matter-energy area. From *6. Producer* leads an information loop to *15. Decoder* which marks the further conveying of reports of scientific activities to *16. Associator***.** The loop in question has earlier been in connection with *15. Decoder* when conveying certain questions to this subsystem. So, this loop is marking two functions, conveying questions and scientific information. It can also be noted that a line is drawn from *6. Producer* to *8. Extruder* which takes care of wastes and takes it out of the system. It can be added that the flows of matter-energy conveyed by *4. Distributor* are marked by the thick line within the matter-energy section in the diagram.

C. Time Flow

14. Timer indicates the timetables for certain scientific projects which are destined for time planning, for example, certain projects within the social sciences and certain research projects within the natural sciences connected to matter-energy flows. For this reason the symbol of timer is connecting the whole system, both the information section and the matter-energy section. This is indicated by the connection lines to *11. Input transducer* and *3. Ingestor* within the matter-energy section.

Special Notations

When studying Figure 19, the question can arise why the functions of *1. Reproducer* and *2. Boundary* are not mentioned. However, the reasons for omitting these functions are that certainly these functions are fundamental to the system, but these functions are not involved in the internal functions of the system itself. Therefore it is possible to say that it is a question of what can be called background functions which back up the system like Atlas in the myth, but which are not involved in the internal functions of the system. Remaining subsystems not marked or commented on are *9. Motor* and *10. Supporter*. However, concerning both of these subsystems the community of science is presumed to rely on systems out of the community as has been mentioned earlier in the presentation of the tasks of the critical subsystems. In this context it also ought to be noted (which is a well known fact) that the exemplified community of science is not a model of an association organized as a special unit but is split up into several different organizational units (faculties and institutions) where each branch of science may have organized its administration in its own manner. That means that the critical subsystems have to be applied to each such organizational unit and that might mean that certain units have to rely on subsystems in near related units within the community of science or in the surrounding society.

4.5. Diagnosing the Organization

Some Words of Introduction

Since the critical subsystems' positions and functions are mapped it is necessary to an investigate their structures. Here it can be of value to look at a questionnaire based on an article in the journal Behavioral Science by Robert Taormina, researcher of the National University of Singapore (Taormina, 1991. vol. 36). This questionnaire has been used when solving the information problems of a Japanese multinational company, a leading manufacturer of control components, electronic transfer systems, medical equipment, information systems and office information systems. However, this questionnaire is limited to encompass only the information processing subsystems. Taormina's motivation for this is: "The reasons for this limitation are pragmatic ones. First, all of the problems in the case described in this paper are presented in the information processing subsystems. Additionally, to make the application of the living systems theory more understandable and to make the diagnostic device more managable and user-friendly, necessitated restricting its length. Also to do this successfully required concentration on only the most important subsystems."

Taorminas' questionnaire has, however, been modified by the author of this report. Thus the letter-indication of the critical subsystems has been replaced by the valid digit numbering. Taormina has also put "what is the recommended structures?" before "what is the existing structure?" The author of this article has chosen to reverse both of these questions because it seems most reasonable to first give an account of the present organization considering the critical subsystems' positions and functions, and thereafter give a picture of the recommended organization. Furthermore, Taormina's questionnaire of the information subsystems has been completed to include all the subsystems. A complete questionnaire with the starting point from that of Taormina follows below. This questionnaire has earlier been published in the British journal, *Systems Research and Information Science*, 1994, vol. 6.

Because the author of this article mostly has worked with problems of municipal organizations, foremost the City of Stockholm (capital of Sweden), the following questionnaire is biased towards municipal problems. Nevertheless, this questionnaire can serve as a starting point for investigations of organizations within the community of science. Of course, here it can be a question of adjustments considering the conditions, structures and functions of the organization respectively. However, even scientific researchers are as individuals members of some municipality and consumers of municipal services of different kinds. Moreover, the scientific institutions depend upon certain municipal services, for example, services from water, sewage, and transportation systems. Therefore it can be of a certain interest even for a community of science to have a look at this questionnaire even if it is biased towards a municipal organization and needs some adjustments when applied to a community of science.

An Example of a Questionnaire of Critical Subsystems

A. Information Processing Subsystems

11. Input Transducer

1. Structure

a) What is the existing structure?
b) What is the recommended structure?

2. Process

a) What is the existing process?
b) What is the recommended process?

3. Pathologies

a) Is the incoming information adequately sensed?
b) Is the information being accurately changed?
c) Is the information being efficiently transmitted?

12. Internal Transducer

1. Structure*
2. Process*
3. Pathologies

a) Is the operations information being adequately sensed?
b) Is the operations information being accurately changed?

c) Is the operations information being efficiently transmitted?

* The structure and process questions are repeated for every subsystem.

13. Channel and Net

1. Structure

2. Process

3. Pathologies

a) Are the channels open?

b) Are the channels operating efficiently/effectively?

i) Can the information be transmitted?

ii) Is the information being purposely withheld?

iii) Is the information being deleted?

iv) Are the messages being delayed?

v) Are messages being distorted?

vi) Is appropriate information being transmitted?

vii) Is conflicting information being transmitted?

viii) Is the information being purposely avoided?

14. Timer

1. Structure

2. Process

3. Pathologies

a) Can information about the start of processing be sent?

b) Can information about the stopping of processing be sent?

c) Can information about the rate alterations be sent?

d) Can information about the change of phase in processing be sent?

15. Decoder

1.Structure

2. Process

3. Pathologies

a) Is the information being decoded (explained)?

b) Is the information being decoded correctly?

c) Is the information being decoded completely?

16. Associator

1. Structure

2. Process

3. Pathologies

a) Are there feedback mechanisms for information to be associated?

b) Is the information being associated?

c) Is the information being correctly associated?
d) Is the information being associated adequately?

17. Memory

1. Structure
2. Process
3. Pathologies

a) Is the information correctly read into storage?
b) Is the information being maintained in storage?
c) Is the information being appropriately altered (updated)?
d) Is the information being appropriately deleted?
e) Is the information being appropriately retrieved?

18. Decider

1. Structure
2. Process
3. Pathologies

a) Are problems/opportunities being recognized?
b) Are problems/opportunities being defined?
c) Are alternative options being generated?
d) Is information on the alternatives being gathered?
e) Are the alternatives being evaluated simultaneously?
f) Is the optimal alternative being selected?
g) Is the selected alternative being correctly implemented?
h) Is the effectiveness of the implementation being evaluated?
i) Is there a feedback mechanism for re-evaluating the problem?

19. Encoder

1. Structure
2. Process
3. Pathologies

a) Is the information being encoded/altered/translated?
b) Is the information being appropriately encoded/altered/translated?
c) Is the information being completely encoded/altered/translated?

20. Output Transducer

1. Structure
2. Process
3. Pathologies

a) Is the outgoing information being adequately transmitted?
b) Is the outgoing information being accurately transmitted?

c) Is the outgoing information being efficiently transmitted?

B. Matter-Energy Processing Subsystems

Even if the questionnaire above is of great importance it cannot go without also saying that the matter-energy processing subsystems have a most important role during the mapping of structures, processes and pathologies in an organization. Therefore it seems necessary to continue the questionnaire about the critical subsystems. However, before going on completing the questionnaire it can be of a certain interest to give a brief definition of the meaning of the judicial status of a Swedish municipality of which the following part of the questionnaire is based: " A municipality consists of a gathering of people living within a certain area of land with boundaries determined by the Government. Each Swedish resident registered with the local tax administration shall be a member of a municipality. Thus the membership of a municipality is compulsory. The municipality has the legal right to act, that is, the right to make binding agreements, the right to free disposal of its property, and to suit and plea in court. Besides the managing of its own affairs the municipality has the duty to carry out certain matters of the State, e.g., the procuration of governmental loans and subsidies in accordance with the housing policy. Here is a question of matters which would otherwise be carried out by governmental district offices. A municipality can for this reason also be seen as a governmental administrative authority." (Rasegård, 1992, p.2).

The matter-energy subsystems can be classified in three groups, namely, production preparation, production and delivery, and supporting subsystems.

Production Preparation Subsystems

3. Ingestor

1. Structure

2. Process

3. Pathologies

a) Are the provisions of material, vehicles, machines and other necessities being satisfied?

b) Is the recruiting of personnel being done considering necessary qualifications (education, practical experience and will of cooperation)?

c) Is any feature of nepotism being distinguished?

d) Are the purchases being founded on offers where all suppliers are equally treated and none is favored?

4. Distributor

1. Structure

2. Process

3. Pathologies

a) Are the transport plans of necessities being rational?

b) Are the transport means suitable?
c) Are the deliveries not being unnecessarily delayed to different receivers?

7. Matter-Energy Storage

1. Structure
2. Process
3. Pathologies

a) Is the book-keeping of the storage articles being done adequately?
b) Is the storage area being used in a rational way?
c) Is the storage area suitable for storing different necessities?
d) Are not current articles removed continuously or an important quantity kept in store?
e) Is the storing being equal to meet demands of fast delivery?
f) Are the costs of keeping storage normal?
g) Is the accounting plan for storage rational?

Production and Delivery Subsystems

5. Converter

1. Structure
2. Process
3. Pathologies

a) Is the converting being made in an effective way in congruence with the demands of the producer?
b) Is it possible to decrease the demands of converting by using pre-adjusted or refined necessities?

6. Producer

1. Structure
2. Process
3. Pathologies

a) Is the production done in a cost effective way?
b) Is the production being adjusted to the customers' actual demands?
c) Is the production apparatus rational considering the technical standards of machines, other facilities and premises?
d) Are certain bottlenecks being distinguished in the production apparatus and, if so, how can these be eliminated?

8. Extruder

1. Structure
2. Process

3. Pathologies

a) Is the extruder function being done in a cost effective way?
b) Is the extruder function capable of working in a satisfying way considering construction and equipment?
c) Is the cleaning and removal or wastes being done considering the environmental demands?

Supporting Subsystems

9. Motor

1. Structure
2. Process
3. Pathologies

a) Are the timetables for personnel transports being co-ordinated considering transport means between different parts of the municipality?
b) Is the air pollution from the transport means being considered in a suitable way.
c) Is the transport net equal to the demands?
d) Are the benefits of different collective traffic means being evaluated in relation to each other (trolley buses, trams, buses, taxi cabs, under-grounds)?

10. Supporter

1. Structure
2. Process
3. Pathologies

a) Are the municipality's traffic ways being maintained properly?
b) Does the main traffic net require re-structuring?
c) Are the plans of locating housings, industries, recreation grounds and communication centers adequate?

C. Information and Matter-Energy Processing Subsystems

1. Reproducer

1. Structure
2. Process
3. Pathologies

a) Is the composition of people normal?
b) Is an occuring distortion of age-classes being caused by political decisions of the municipality (e.g.,high taxes and charges and not so good services?
c) Is the distortion of age-classes being caused by political decisions of the government (e.g. lessening of subsidies to vital services or defects in the central planning policy)?

d) Are all opportunities of applying for governmental subsidiaries being considered?
e) Are conditions for starting new enterprises favorable?
f) Is there an imbalance between male and female opportunities for jobs?

2. Boundary

1. Structure
2. Process
3. Pathologies

a) Is the municipal geographical boundary adequate considering topographic, social and economic demands?
b) Is the information threshold so high that certain important information risk being neglected (e.g. certain ecological information).
c) Is the limit of salaries and wages so low that it is difficult to recruit qualified personnel to municipal services?
d) Are the communication and transport means to and from the municipality satisfactory?
e) Is the cooperation with neighbourhood municipalities normal?

D. Some Comments on the Process Concept of Certain Subsystems in the Questionnaire

1. Reproducer

Even if it can seem a bit peculiar to speak of the *process* of this subsystem, it can not go without saying that it can be a question of investigating the effectiveness of the routines of taxation, combating economic and other criminality and applying for state subsidies. The reproducer, as known, consists of the municipal citizens, who besides the pure physical reproduction are responsible for the economic reproduction. In order to facilitate the physical reproduction it is also important to investigate that plans for maintenance re- and new investments of maternity hospitals, day-care centers and schools are up-to-date. Here the objection can arise that the tasks above are connected to other subsystems. Nevertheless, these tasks, even when carried out by components of other subsystems are of initial meaning for the reproducer function.

2. Boundary

Maybe it can seem rather surprising to speak about the concept *process* related to boundary, but the following examples are aimed to shed light on certain problems connected to the subsystem in question.

Matter-Energy Boundary

a) Watching for signs emanating from air and water pollution sources situated out of the municipality. This task usually is performed by personnel at the health- and environment authorities.

b) Tests and investigations of artifacts and necessities before these are used or taken into storage. These tasks are carried out by storage personnel and experts from the user-field.

Information Boundary

a) Watching that the municipality's competence is not distorted by governmental laws submitted to the municipal authorities for pronouncement. This task is carried out by the municipality's judicial experts.
b) Watching that decisions by the municipality's deciders are not in conflict with the municipality's legal competence. This task is performed by the municipality's judicial experts and auditors.

Even if certain components of the boundary process at the same time can be included in other subsystems, e.g. input transducer and ingestor, it is obvious that a certain part of it, as the examples above illustrate, is of a pure boundary meaning.

10. Supporter

Here it can seem strange to mention the concept *process* in as much as the subsystem in question does not contain living components. But in this case it is a question of what the producer does when maintaining and re- and new investing of the infrastructure. Thus, it can be a question if the producer to a sufficient degree is capable of maintaining and developing the infrastructure or, in other words, the degree of how the municipality can manage these tasks in its own production, and how much must be done by external entrepreneurs. Besides personnel connected to the producer, personnel from other subsystems, for example, planning teams connected to the associator, are involved in the supporter process, even if the subsystem in itself does not contain living components.

4. 6. Examples of Malfunctions

Malfunctions Related to Financial Policy

Perhaps it can seem a little curious speaking about financial questions under the subheading above. However, the critical subsystems consist of researchers with assisting personnel and their artifacts, and therefore it seems self-evident to connect the subsystems to financial policy.

It may seem a bit strange that the scientific paradigms can be influenced by the political environment, but the political environment represented by government and its scientific advisers exerts its influence through financial means. Thus, research that is judged as out of a certain paradigm does not have any opportunity to receive grants or other financial support. In this connection it can be mentioned that establishing *systems science* as an acknowledged scientific discipline has sometimes been met with resistance among the spokesmen of the older disciplines. James Grier Miller has in connection with this resistance made the following remarks when he, as President of the University of Louisville, was feeling an aversion to give his discipline a fair chance according to the other established academic disciplines: "It is fortunate that the good Lord created the Universe exactly according to the academic disciplines." And he continues: "Every academic discipline has its own jargon - and there is no Rosetta stone." (Samuelson, Kjell, 1982). As another example of downgrading the discipline in question can be mentioned that the interdisciplinary science *informatics and systems science* in Sweden has met with difficulties in receiving financial means compared to other disciplines. However, it can not go without saying that the preserve- thinking of scientists from other disciplines plays an important role. That fact can have a negative consequence for certain research objects which in the eyes of autocratic committees of scientists (who are influenced by politicians) can be regarded as odd objects not worth spending money on. Here it can stand to reason to refer to the following statement by Hanbury Brown: "Admittedly, the total amount which is spent on basic research is a matter to be decided by the government of a country, but the detailed disbursement of that money can only be done effectively with the help of a group of scientists acting on behalf of the scientific community. If this group is to be up-to-date, they must be working scientists, and if they are to ensure that science advances on a wide frontier, they must be drawn from as wide a range of scientific disciplines as possible." (Brown, 1986, p. 119).

In this context it can also be of interest to refer to the following citation about the situation of research funding in the U.S.A.: "A critical element of future research is of course research funding. In the past three and a half decades, we have seen the unparalleled dominance of the Federal government in paying for research in the United States – starting, it seems, as a response to early consternation over "Sputnik" and subsequent Soviet advances in technology. That dominance of the Federal government in research support persists, and will probably continue, subject to some fluctuations and to the whims and attitudes of the administration in power. This means that decisions about funding for research are being made and will be made by politicians (aided by whatever scientific advisors they choose to listen to). Some observers view this concentration of control with alarm, objecting to the narrowness of vision and the mission orientation of government funding agencies. Others see government funding as the only solution to the needs of 'big science' and to the uncertainties of private sector funding."(Sinderman, Carl J., 1987, pp. 250-251). In this connection it can be of a certain interest to cite another statement by the same author: "Accepting the likelihood of the continuing presence of the government in decisions about research grants, new scientists would be well advised to seize opportunities to understand the complex techniques of decision-making and control that have developed in many funding agencies and to participate with great vigor in any scientific review and advisory procedures made available by those agencies. A relatively new role for scientists, which has developed in response to government dominance, was described earlier in this book as that of the 'scientist-politician'- a credible professional who has learned to deal effectively with systems of control and who can influence those systems to move in the best interest of the science community. For these scientists with the aptitude and interest, this role gives every indication of being more important in the future." (Sinderman, 1987, p. 251).

To sum up it can be said that in all scientific disciplines , it is of vital importance that the spokesmen of the respective discipline have the will and ability of driving forceful and enlightening information about actual research objects. And here the information to the outside world in the form of press reports, radio and television programs and open lectures (popular science) play an important role (cf the functions of the critical subsystems 19-20). Certainly it can be argued that research within the area of military defense is of a secret nature and cannot be the object of information made public. But here the researchers have to convince the military decision makers and connected politicians of the necessity of giving financial support to what the researchers consider to be important topics. However, even if the military research of pacifists is regarded as being of a purely destructive nature it cannot go without saying that even this type of research can have certain positive spin-off effects even on the civil research objects, for example, space technology.

Malfunctions Related To The System

When observing the functions of the critical subsystems in an organization, it follows that an analytical study of how each subsystem performs its duty. Therefore it gives the following examples of malfunctions related to individuals and artifacts.

Information Section

11. Input transducer can have missed important information. It can be that a scientific attaché at an embassy abroad neglects to report home some scientific progress, or a book reviewer has not observed an interesting book. Another mishap can be that certain researchers are unwilling to give complete information of their findings because they think that other researchers will parasite their work.

12. Internal transducer. Here, for example, an outdated accounting system can cause trouble in obtaining a relevant economic information for using grants and other financial means.

13. Channel and net. A serious source of malfunction can be that the computed data net can risk infection by data viruses which can destroy important information. The same can be said of information stored in the computed data bases.

14. Timer. The malfunctions here can be that clocks constructed to measure precision of scientific experiments are not reliable depending on the carelessness of personnel who have the responsibility of attending. Thus, failures of artifacts connected to the timer can have the most serious consequences for scientific research within certain disciplines. Another factor which is involved in the timer function is setting of timetables for research objects. This can under unlucky circumstances lead to financial difficulties when a project takes a longer time than originally calculated.

15. Decoder. Here it can be so that translators of foreign literature can misunderstand some passages and shaded expressions which can lead to certain phrases receiving a meaning other than the one the author has intended. That malfunction, however, is not so impending concerning the languages of the leading industrial nations in the West but can be a serious factor when it comes to more exotic languages, for example, Chinese and Arabic. The malfunctions of the decoder can also be manifested in lack of ability of making clear the rules based on laws and ordinances which are an important part of administration of the community of science. The scientists, as known, are not as a rule interested or competent to interpret the actual contents of these laws and ordinances .

16. Associator. This subsystem contains advisers who give advice to the political decision makers. Therefore it is of essential value that these advisers represent necessary competence and are free from obligations to both institutions connected to the community of science as well as political and commercial interests. Another question of general interest and of great importance is that leaders of scientific activities are selected only by considering real merits of scientific education and training and competence of research. Consequently, the recruiting ought not be done, for example, for reasons of politics and gender.

17. Memory. Here it can be remarked that the danger is that outdated information is not weeded out but is accumulating in the data bases. The same can be said about manually operated libraries and archives. However, as mentioned in another connection, the weeding out of undesirable information must be done carefully and supervised by experts of the scientific discipline in question.

18. Decider. As mentioned earlier, this function, can be divided into two levels, the political deciders and the scientific deciders. However, it would be noted that the scientific deciders do not suggest that the latter type of deciders are making special scientific decisions. No, these deciders at universities and other scientific institutions are the persons who have the overlapping responsibility for appointing researchers and other personnel for the area of education and research including recruiting of new students. The malfunctions can be, as have been said in an earlier context, that the political deciders have not received enough information by the advisers in the associator function and that the scientific deciders have been infiltrated by politicians without enough competence and will to grip new information and development. The same can be said of certain political deciders with a background as scientists. It can also be that the scientific deciders (especially found in the social sciences) are too trapped in a certain political paradigm which hinders a free investigation of different possibilities.

19.Encoder and 20 Output transducer . As mentioned earlier, the functions of both of these subsystems can be coupled together. The malfunction here can be that the subsystems in question cannot fully understand enough to be able to give easily available and comprehensible information to politicians and common people. If so, people can get the idea that scientists are some form of clan living isolated in an "ebony tower."

Matter-Energy Section

3. Ingestor can by lack of sufficient financial means be forced to purchase cheap artifacts and material of not so good quality. But even here lack of sufficient competence can be serious due to the fact that the salaries are too low compared with the private enterprises .This is also a problem when it comes to recruit qualified researchers and other personnel in key positions. Another factor of malfunctions can be that personnel involved in purchasing connected to the ingestor has neglected to keep up with the last development of artifacts and materials and that can depend on ignorance of the input transducer when conveying the latest information. It can also be that a purchase department has strong connections to an old and influential supplier which it has difficulties breaking both for formal and personal reasons.

4. Distributor can be exposed to malfunctions if its own transport means are insufficient for rapid transports between, for example, different laboratories for transports of test materials, the public transport means are too slow or the timetables are not fitting the demands of urgent transports or that certain types of transport are too expensive for a limited budget. The latter is probably not a problem of great importance in the developed industrial nations but it can not be overlooked in lesser developed nations.

5. Converter. Here malfunctions can be caused by inferior artifacts for refining and adjusting the necessities used in the producer function and lack of necessary competence of the personnel, for example, instrument makers with insufficient training. These malfunctions can have their roots in too limited financial frames, but also in neglect of the responsible authorities. In other words, the spokesmen of a certain discipline may have not had the ability of convincing the respective authority or its representative about the real need.

6. Producer can , for example, work in a below average way because of inferior artifacts in the form of scientific utilities. It can be caused by financial means being too meager to permit purchase of high class utilities or the competence of the personnel is too low as a consequence of too limited salary frames. The consequence will be a poor quality of the research connected to laboratories and other working-places of natural sciences.

7. Matter-energy storage . Here it can be that poor book keeping can disturb the supplies of necessities by the ingestor . Another source of mishap is that the storing is not equal to meet demands of fast delivery.

8. Extruder. Malfunctions can be caused here by carelessness of the cleaning personnel but also if extruder functions carried out by external entrepreneurs are not done in a satisfying way.

9. Motor. Here it can be said that this subsystem can be regarded as not belonging to the community of science . Nevertheless, representatives of the community of science can be said to have a certain responsibility for making proposals of improved timetables and transport means to the traffic authorities.

10. Supporter. Concerning this subsystem it can not be a question of malfunctions because it does not in itself contain living components. The malfunctions here are related to the society's subsystem producer whose duties are to maintain the buildings connected to scientific education and research and other constructions in a proper way. However, if a scientific institution of education and research is of a private nature it can be the producer function of the organization respectively that has the duties mentioned in the foregoing or a private entrepreneur out of the organization in question.

The exemplified malfunctions here which can arise in the critical subsystems can be traced, and the consequences analyzed in a more precise way when going through a scientific association and the duties have been located to the different subsystems respectively. Then it also will be easier to decide what sort of competence would be demanded of the individuals connected to the respective subsystem. In this connection it will also be a mapping of communicative ways between different subsystems and how they are cooperating. Because the community of science, as mentioned earlier, is split up into many different disciplines and faculties it can be presumed that a complete set of subsystems are not to be found within each discipline, but that we here see that certain disciplines have to rely on subsystems in other disciplines – or subsystems in the society outside of the community of science.

Chapter 5

ECONOMIC STRUCTURING

5.1. INTRODUCTION

Even, as shown in an earlier context, this report is mainly related to certain problems of the community of science, it cannot go without saying that examples of municipal planning also can be of interest for organizations within the community of science because the universities and other institutions of scientific research are situated in some municipality.

Therefore the authorities of the community of science , boards of universities and other scientific institutions can have a certain interest to look at what a municipality is planning for. Furthermore, an example like the following can give impulses even for planning within suitable parts of the community of science, foremost the administrative body and other organs within the community occupied with practical works. Of course, it can be necessary with certain modifications related to the structure of the respective organizational unit in relation to the example below. Nevertheless, at least it can be something to think about.

To begin with it can stand to reason to mention some common words on the meaning of planning. In the course of recent decades we have witnessed various efforts to apply what have been considered to be effective tools in improving planning and structuring of an organization. In considering these tools we have to ask ourselves certain fundamental questions, such as: What is the innermost purpose of what we are planning for? What is the desired achievement ? How can we predict that this achievement corresponds to the needs of customers and the environment ? Do the deciders have the necessary versatility to cope with changing problem views in a more long-sighted perspective or are they trapped in a petrified old paradigm? Will the decisions process be delegated to a lower level of the organization and, if so, how?

Attempts to solve these and related questions have been made by various methods. We have, for example, been acquainted with the PPB (planning, programming, budgeting) system as a tool to establish a more effective control of costs and their spending in various fields of activities. We have been presented with another method of budgeting against lack of effectivity but also as a means of delegating decisions from higher to lower levels in an organization, namely, zero-base budgeting. We have also been familiar with operational research as a way to make rational decisions. As a consequence of operational research we have at least attempted to apply ideas of cybernetics in organization development. The latter has supported the creation of the interdisciplinary discipline that we know under the name of systems science – in Sweden named "informatics and systems science".

5.2. Implementing the Critical Subsystems Combined with Zero-Base Budgeting

A problem when analyzing an organization according to the ideas of systems science in order to form a new structure is to find the most practical way to do this work .One way to approach this problem can be to combine the ideas mentioned in chapter 4 with the ideas of zero-base budgeting and in some way merge together the former ideas with the latter. Zero-base budgeting is a method for budgeting, originally based on Peter A. Pyhrr's book of the same name from the time he was working at Texas Instruments (Phyrr, 1973). This type of budgeting has, among others, been used by former U.S. President Jimmy Carter when he was governor of the state of Georgia.

Zero-base budgeting has as a starting point the lowest level of budgeting responsibility within an organization and consequently presumes that the responsibility of budgeting is delegated to small sections of the organization. It can, for example, be a fraction of a section of a department or other small unit. This unit then has to work out decision packages for different levels of its activity at the same time that it presents the goals and costs to reach these goals. Decision packages can be defined as alternative calculus for a certain activity. Because zero-base budgeting divides the organization into a number of small independent budgeting sections, it seems very suitable to combine this method with an organizational analysis according to the principles of critical subsystems. A simple example which illustrates how zero-base budgeting would be combined with the previously-mentioned principles is shown in Figure 12.

Here it is presumed that a department of an organization is divided into bureaus which in turn are divided into sections. The decision packages in the example are only worked out on three levels , namely, increased, present and minimum activity volumes, even if in practice it can be desirable to show the consequences of, for example, a reduced volume below the minimum level for a normal activity. As the figure indicates, each package is marked with a two-digit code (xx) which indicates the subsystem in question. Therefore it can be a matter of, for example, *11. Input transducer* and *15. Decoder,* etc. When these decision packages are treated the management of the organization has the possibility of obtaining at least a rough draft of the costs destined to each subsystem besides those spent on various activities, from the smallest to the largest, and their outcomes. The two-digit code is intended to accompany the decision packages through the whole budgeting procedure.

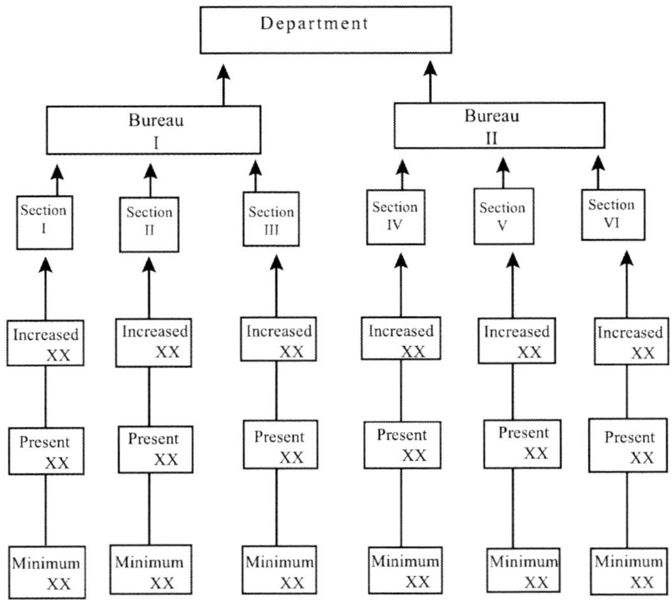

Figure 12. The critical subsystems indicated in decision packages (from Rasegård, 1986).

5.3. DEFINITIONS OF CONCEPTS CONNECTED TO ZERO-BASE BUDGETING

The previous section had advocated using zero-base budgeting as a tool when analyzing an organization according to ideas of systems science, mainly represented by the critical subsystems. Therefore it seems necessary to give definitions of the key-concepts involved in the mentioned method of budgeting according to the following:

Zero-base budgeting is a method of budgeting which presumes an unconditional examination of a program section to the lowest area of responsibility, and where the activity within each activity-section is investigated yearly from the base in respect to if it has a meaningful function within the organization. The method is based on prerequisites of alternative budget calculus in the form of so-called decision packages.

Decision-package is a budget calculus which contains a statement of the expected result, the purpose of the activity, its costs, necessary personnel and, in certain cases, proposals of alternative solutions concerning spending of resources and the level of ambition to carry on the activity, for example, minimum levels, present levels and increased levels. Decision packages below shall also contain an estimation of the consequences of abolishing a certain part of resource spending with consideration given to the present level of the respective decision package.

Package level is the level of ambition which is represented by the goal of the activity considered to in the respective package's indicated input and output.

Minimum level is the level of input and output below which it is unrealistic to carry out any activity at all.

Present level is the level of input and output of an activity which presumes that it can be carried on in a normal way according to the decision makers' intentions considering the level of service and work load. The budget calculus on the present level shall in addition to adding for increased costs caused by inflation and other factors, shall also add for possible enlarging of the activity that lies within the frame of the present level, for example, increased costs as a consequence of enlarging of some unit´s activity.

Reduced present level is the level of input and output which means a certain reduced activity compared to the present level, however, without bringing down the activity to the minimum level.

Increased level is the level of input and output which means an increased activity with improved service and standard, for example, more resources for research in the form of personnel and equipment. It can also mean increased input and unchanged output but improved service and standard , in other words higher quality.

Ideal level is the level of input, output and standard desirable if no restrictions in the size of input and hindrances of an organizational nature would be considered . Packages of this level can be regarded as instruments for brainstorming in order to promote and develop the personnel´s creativity. In contradiction to the underlying package levels a final ranking and judging is not demanded. The packages in question need consequently not have room within the cost-roof determined for the actual fiscal year or a certain planning year.

Cost-roof is the level of costs above which an activity is not permitted to be carried on. The cost-roof can be an agency´s total resources according to the superior boards directive and below that, the appropriations which are directly connected to the program section respectively and between which a subordinated board can make priorities.

Head-label is the common denotation of a main division of decision packages into:

- operational packages and
- development packages.

Operational packages is a summary denotation of these decision packages which are involved in the operational activities.

Development packages are packages which are aimed at an accounting for input and goal-setting considering development of a certain type of project out of the operational activities, for example, development of new methods for using a new computed- based system for research. Development can as well as operational packages be worked out on different levels of ambition.

Basic packages are decision packages which give account for the main alternative on different levels of ambition, for example, minimum, present, and increased level. Besides the basic packages, alternative packages can be worked out considering different types of input in order to obtain the same goal-setting accounted for in the basic package.

Alternative packages are decision packages which are worked out as alternatives to a package on the same level of ambition in order to give another solution for the types of input for the same activity, for example, employing external entrepreneurs for a certain part of the organization, for example, administrative works.

Operational classification refers to the basic meaning of the decision packages concerning accounting for resource spending and outcome of a certain delimited activity on different levels of ambition, and within certain cases alternative solutions. The decision packages can for this reason be classified in basic packages and alternative packages.

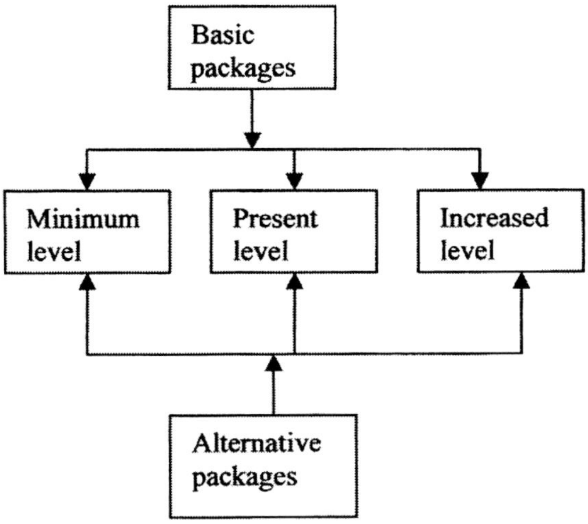

Figure 13. Operational classification (from Rasegård, 1992).

Functional classification is aimed to make clear the different functions of the decision packages within both the main groups *operational packages* and *development packages.* Within both these main groups the decisions packages can be classified

- according to the legal function into
    - special packages and
    - municipal packages
- according to the planning time-horizon into
    - long-term packages and
    - short-term packages
- according to the way of financing into
    - self-financing packages and
    - taxes-financing packages

Special packages are decision packages concerning activities which to a great deal are regulated by ordinances given by the State, for example, health- and environment –protection.

Municipal packages are decision packages concerning activities where the municipality has unrestricted influence, for example, athletic activities.

Long-term packages are decision packages of which the time horizon is equal to those of the long-term planning and consequently are extended for more than one year. The long-term packages, can in the same way as the packages in the yearly budget, be divided into operational respective development packages and be worked out on different levels of ambition.

Short-term packages are decision packages of which the time horizon is equal to the yearly budget.

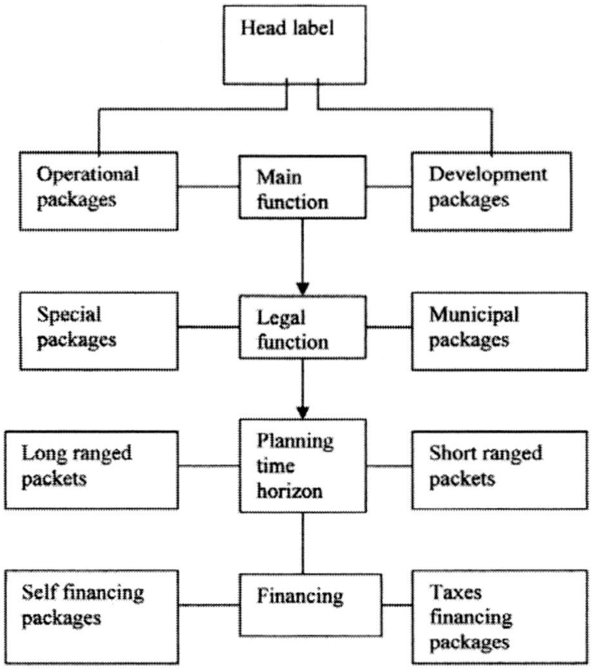

Figure 14. Function classification (from Rasegård, 1992)

Self-financing packages are decision packages where the costs are fully covered by revenues of the activity represented by the package. If a decision package shall be denoted as self-financing, the revenue consequently must be delimited and directly destined to the section of activity to which the package belongs, for example, certain packages which are parts of the community´s activities financed by tariffs and taxes.

Taxes-financing packages are decision packages which in full or in part are financed by taxes.

Program-section is a delimited section within an agency and which is directly connected to an appropriation in the budget , for example, a larger establishment of athletics.

Activity-section is a part of a program-section of which the economic frame, a partial appropriation, is determined by a respective divisional board, for example, the part of an appropriation directed to a special type of athletics.

Package-section is within an activity delimited section of responsibility which either can be designed to fulfil a certain function of responsibility or refer to a certain part of an activity, for example, a day-care center. The package-section is in the decision package denoted as activity.

Function is a part of a package-section including a certain type or certain group of tasks and which in itself does not need to be limited concerning responsibility, for example, maintenance function or administrative function. In what degree a function in respect of responsibility is clearly delimited, is equal to a section of responsibility under the concept of package-section.

Purpose description is a summarized description of the purpose of an activity. Can be compared to a newspaper placard.

Operational goal of activity is the desired outcome of an activity expressed in operational terms related to the output, for example, the number of school meals produced. It can be common to more than one decision package and is in this respect superior to the evaluation measures on lower levels which are connected to respective decision package.

Work efficiency (productivity) indicates in what degree one within a section of responsibility is able to economize the volume of work connected to the goal-setting within the respective activity-section. Usually expressed in costs per produced unit.

Resource type is the type of costs specified in the organization´s accounting plan.

Resource list is a specification of the different types of resources which an agency uses in its activity process.

Resource-checking is a checking within the economic frame which is done in connection to the final ranking of the decision packages.

Ranking means putting in order of precedence the decision packages within a section of responsibility with consideration to priority. Ranking can be done on different levels of decision within an agency where each decision package is an object of reconsideration.

Decision unit is the organizational unit responsible for budgeting which is ranking the different decision packages according to priority. Thus, decision units are to be found at different levels in the municipal organization.

Absolute priority is applicable to decision packages which concern activities regulated by ordinances of the State, so-called special-packages. Thus, these packages are given the highest priority in the ranking procedure.

Cut-off level is the level of costs (cost-roof) where delimitation is done between packages that shall be transformed into a budget and packages that shall be rejected.

Cut-off packages are decision packages which at the final ranking have been found lying above the cost-roof and therefore are rejected.

Ranking plan is the plan for how the ranking procedure shall be carried on in an agency. The plan indicates the different levels on which the ranking shall be done, the order of negotiations with the representatives of the personnel, the officials who shall take part in the ranking procedure, and a time planning for the procedure indicating a deadline.

Ranking list is the list in which the different decision packages are put in order of precedence. Ranking lists are worked out at every level of decision within an agency.

Ranking record is a record kept at the ranking meeting with reasons for the decision of priorities among decision packages.

Ranking manager is the leader of the team who at every organizational level makes the first judgment of priorities among the various decision packages . The leader of the ranking team has the responsibility that the ranking procedure will occur at the right time and that all of the judgment material is accessible. The leader of the ranking team together with the members of his team is responsible for priorities made at the actual organizational level included in the ranking.

Package manager is the official equipped with authority to exert influence on the activity which the decision package is aimed at. The package manager has the primary responsibility for designing and working out the decision package. As a consequence of the management function he signs all packages within his section of responsibility.

Ranking information means the information about priorities made among decision packages and the reasons for these, which the respective decision unit ought to give to responsible personnel at lower organizational levels. This information can be given formally on ranking records.

Note

It seems necessary to make clear that the definitions above are adjusted to the organization of the special structure of a municipality. Thus, an organization in another public or private sector of society can, in certain cases, have different denotations for these basic concepts of zero-base budgeting. Neither is the description of zero-base budgeting given above a complete account of this form of budgeting , but ought instead, to be regarded as an introduction to a budgeting method which lies on the same line as a system based on ideas of systems science with their roots in cybernetics.

Chapter 6

USING BREAK-THROUGH THINKING IN A PLANNING PROCESS

Another instrument for solving the organizational problems of planning and production is 'break-through thinking' (Nadler and Hibino, 1991). The seven principles that the authors put forward are:

1. *The uniqueness principle*: Each problem is unique and requires a unique solution.
2. *The purpose principle*: focusing on purposes helps strip away non-essential aspects of a problem.
3. *The solution after next principle*: Having a target solution in the future gives direction to near-term solutions and infuses them with larger purposes.
4. *The systems principle*: Every problem is a part of a larger system of problems, and solving one problem inevitably leads to another. Having a clear framework of what elements and dimensions comprise a solution assures its workability and implementation.
5. *The limited information collection principle*: Excessive data-gathering may create an expert in the problem area, but knowing too much about it will probably prevent the discovery of some excellent alternatives.
6. *The people design principle*: Those who will carry out and use the solution should be intimately and continuously involved in its development. Also, in designing for other people , the solution should include only the critical details in order to allow some flexibility to those who must apply the solution.
7. *The betterment timeline principle*: The only way to preserve the vitality of a solution is to build in and then monitor a program of continual change. The sequence of Break-Through Thinking solutions become a bridge to a better future" (Nadler and Hibino, 1991, p. 88).

After mapping the organization, the whole or part of special interest, and the structures and processes of the different subsystems have been analyzed, it is time to apply the seven principles of break-through thinking to current problems of the organization. In this connection it can be worthwhile to refer to a case from the practice described by the problem. Ahari says:

> "Using Breakthrough Thinking gave me insight into the situation. This implied: Do not accept automatically senior members' suggestions or imitate previous solutions. I did not try to find out what the 'problem' was. Instead I tried to understand the purpose of our work. I saw the task as unique (the first principle of Break-Through Thinking) and I tried explicitly to find out the purpose of our work (the second principle of Break-Through Thinking). Using the Limited Information collection Principle (the fifth principle of Break-Through Thinking) I reviewed the material and found that this information (the explicit statement of the purpose) did not exist in any document. My first suggestion was that the group should try to find out the purpose of our work and write it down explicitly and clearly so everyone could understand it. This could help us to reach an agreement described in the second principle of Break-Through Thinking. The last step was to keep the customer satisfied with the product and services of the company. Between these stages I had created other purposes , one of which was the following: 'To create information/documentation which makes the task of product handling easier for the customer'. This formulation was a dramatic change and it opened other possibilities and generated some ideas about which goals to seek. There was a better agreement on this formulation . Different sources of information from user groups within customers' organizations and within the company could be used to collect customer/user preferences, comments and suggestions concerning both the new and the current (similar) documentation" (Ahari, 1992, pp. 466-467).

Now it can be of interest to explore whether break-through thinking can have any meaning in a public organization. As an example, a municipal organization is selected because of the authors' long experience of investigations of questions concerning municipal planning and economic management systems. The main problem is how to improve the services to the municipal citizens.

Considering the first principle, the *uniqueness principle,* it can be said that this problem is embedded in a larger system, namely, the system of society ruled by parliament and the government. In this larger system the organizational system of municipality has a function as a distributor of the overlapping welfare decided by State authority. That means that a certain part of the social welfare in a municipality depends on the laws and ordinances issued by the government. It also means that a part of the municipal income is emanating from the State and is bound to certain conditions. And that means in turn that the municipal welfare policy is partly dependent on decisions within a larger system; therefore the uniqueness of the problem is limited to the type of services which the municipality has unconditional possibilities to influence. Thus, in the following example we will investigate which possibilities the municipality can have to improve the service to the citizens in the form of better traffic conditions. This is also a problem which can be of interest to a community of science.

The purpose principle can be exemplified by the following: to improve transport of people by collective traffic means within the municipality. The purpose behind that purpose can be a program for diminishing exhaust from motor vehicles by using trams , trolley buses, and underground trains. The purpose behind the purpose of that purpose can be, in a longer perspective, to influence the State to lay main thoroughfares out of the municipality's densely populated areas . A first solution related to the foregoing purpose can be arranging parking spaces for unnecessary cars, private cars, in the outskirts of the city in order to eliminate traffic jams in the inner city. This solution involves improving collective traffic means.

The solution after next principle can then be of help when thinking about the plans of locating housing, recreational and industrial areas. As mentioned above the thoroughfares financed by the State come into the picture. Here the solution after next can be to consider a restructuring of certain areas depending on the future traffic plans presented by the State. That can mean rearranging the planned traffic in certain areas.

The next principle is *the systems principle*, which states that every problem is a part of a larger system. Using the example above it can be pointed out that the municipal problem of how to improve traffic flow is a part of a larger problem of how the State intends to improve the main traffic arteries for which it is responsible – and that can be connected to another problem, namely, the State's intentions for developing different regions and the municipality's region in particular.

The limited information collection principle, the next principle, can be considered as a warning of information overload. Too much information can contribute to overshadowing of other alternative problem solutions. The danger is that the information can be a pile of information focused too much on solving the problem of present traffic intensity and not so much on the other aspects mentioned above. Here it can be worthwhile to cite Peter M. Senge. He says:

"The increasing complexity of today's problems leads many managers to assume that they lack information they need to act effectively. I would suggest that fundamental 'information problems' faced by the managers is not too little information but too much information. What we need are ways to know what is important, what variables to focus on and what can help groups and teams develop a shared understanding."(Senge, 1990, p.128).

The following principle, *the people design principle*, is very important because a condition of a good solution is that the people who continuously are involved in the practical work must take part and give their opinions of a problem's solution. Therefore, in improving the traffic flow it is important to involve representatives of the area concerned and representatives of different traffic organizations and associations. In this context it is important to create an atmosphere of team creativity in order to obtain a synergetic effect.

At last we can consider *the betterment timeline principle.* Going back to what has been said earlier about improving the traffic flow it can be a question to take into consideration conditions for further transfer to electric cars, trolley buses and trams. In this connection it is also important to consider traffic connections with the municipal center and airports and the consequences of changing to a more airborne traffic.

Chapter 7

SUMMING UP THE STRUCTURING AND PLANNING PROBLEMS

7.1. SOME QUESTIONS CONNECTED TO ZERO-BASE BUDGETING IN A PLANNING PROCESS

1. In what way can zero-base budgeting be of help when discovering the critical subsystems in an organization?

Zero-base budgeting is aimed at dividing an organization into the smallest units of responsibility, for example, a secretarial or accounting section within an administrative bureau or a supervisor's section within a bureau for street maintenance . When dividing the organization into these small budgeting units, it will be possible to discover the critical subsystem(s) within the unit in question. This can be said to be an automatic result when mapping each unit's type of work which will be indicated in the budget form.

2. To what degree can zero-base budgeting make it possible to detect how the discovered subsystems cooperate?

In connection with mapping of the budgeting units, it may be necessary to have a look at the individual work descriptions within the unit in question.

Such a step ought to be, in certain cases, completed with personal interviews. During this investigation it will be clear how the functions of the actual subsystem(s) are carried on within a certain budgeting unit and how they cooperate. The author of this report has earlier in another context given an example where an unsatisfactory cooperation caused the municipal organization in question (the city of Stockholm, capital of Sweden, a considerable economic loss).

3. Is it possible to obtain an exact cost accounting for each subsystem within the organization besides the costs of each budgeting unit?

As indicated earlier it ought to be possible to obtain at least a rough draft of the costs destined to each subsystem. It is possible to make an exact picture of the costs if it is possible to make a clear distinction between each individual task related to a certain subsystem. However, in practice an individual can carry out tasks connected to more than one subsystem. In such cases it will be necessary to make an approximation of the rate of work which is connected with each subsystem. But the costs consist not only of salaries and wages with connected overheads; it is also a question of distribution of other overheads to different subsystems. Here it can be a question of finding suitable key norms for this distribution in proportion to the individual rate(s) within each subsystem.

When implementing the ideas of the critical subsystems, however, it is not necessary to work with the intention of bringing about an exact cost accounting for each subsystem during the first year because such work can contribute to the hindrance of the analytical work aimed at developing the organization's working structure. Therefore it seems reasonable, at least during the implementation phase, to be satisfied with rough rules of thumb. These rules of thumb can later be refined step by step when experience is gained about the development of the organization's structure.

4. Earlier a two-digit code was exemplified to indicate the critical subsystems within the organization in connection with zero-base budgeting. However, in the former paragraphs it is hinted that perhaps it can only be a question of rough estimations of costs; therefore, why insert this code which perhaps is not continually used?

It is correct that use of a certain code can cause some trouble when, for example, you have to divide an employee's task and associated costs into more than one subsystem. Furthermore, it can cause trouble when an employee leaves his/her position and is substituted with another who partly has been assigned other tasks as a consequence of exchanging parts of the task between different individuals. But this is a question of reporting in order to maintain an up-to-date work description, which also has a function when it comes to salary and wage negotiations between the employer and the trade unions. However, it can be remarked that if, in a beginning phase, the workload can seem to be to heavy, it is possible that one in certain cases can leave out the use of the code of the critical subsystems in the budgeting and cost accounting process. Nevertheless, space for this code ought to be reserved in the code system in order to make it possible to do special investigations within particularly interesting sections within the organization. It can, for example, be of interest to investigate the degree of involvement of external consultants in the associator function.

Even if the cost accounting in certain cases can be rather schematic, considering the critical subsystems, it ought to be valuable to get a code of these subsystems related to those individuals who are working within the function section of the respective subsystem. Thereby it will be possible to obtain foundations for investigations – for subsystem – for example, if certain individuals have the necessary qualifications corresponding to the demands of both the internal and external milieu. We can here, as an example, choose *11. Input transducer* and *15 Decoder*. Another subsystem which can be of special interest in this context is *17. Memory*, which contains both computed and manual routines. Here it can be valuable to scrutinize the different parts of this subsystem, for example, to perform investigations of the time for access and updating at the different parts of the organization, and look at the degree to which the manually kept diaries and records are made available and kept up-to-date considering actual events. It is also of great importance to investigate that outdated information has been weeded out.

However, it is important to remember that the critical subsystem's role in an organization *a priori* is to form an apparatus for the analyzing of the organization as a functioning living system. Seen from this viewpoint it seems unnecessary to give an account of current performances of these subsystems in every detail connected with decision packages in zero-base budgeting. It can instead, as mentioned in an earlier context, be a question of particular investigations in order to obtain a picture of the functions of the critical subsystems within a certain section of the organization. Finally it is important not to leave out the practical consequences of time, money and workload.

7.2. Certain Critical Elements in a Planning Process

These examples of municipal planning may seem trivial, but they are nevertheless aimed at pointing out some ideas which are involved in breakthrough thinking. As previously mentioned breakthrough thinking directly influences both solving a problem within an activity as well as the activity's relations to its environment. In an organization structured according to the critical subsystems, these subsystems are presumed to be integrated in an activity connected to zero-base budgeting as well as in problem-solving groups within the activity. Beer's different levels of desired outcomes come to expression in the production volume of the activity's performances . Inserting these levels of performance in the planning system can also in turn result in focusing on certain problems of production where breakthrough thinking can contribute to problem solving in the short term as well as in long term planning. When using the seven principles of break-through thinking there is no need, as the authors point out, for a procedure to determine which step has to be used first, but breakthrough thinking will work best if you apply all seven principles all the time. First among equals, however, is *the purpose principle*. The final planning, however, where the result of using the elements pictured in the figure below is documented in the decision packages of zero-base budgeting. Putting what has been said concerning certain planning elements into a framework, Figure 15 can serve as an illustration.

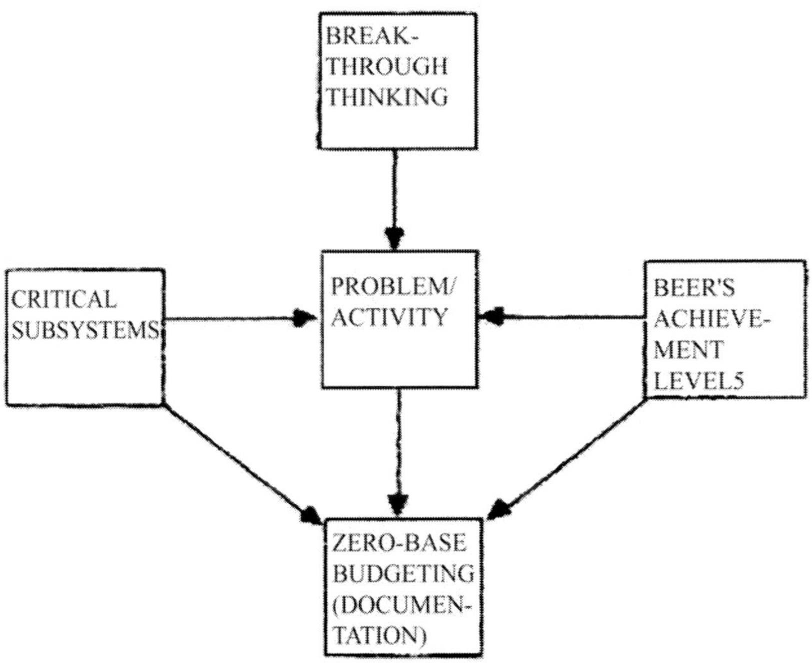

Figure 15. Critical elements in a planning process based on systems science ideas (from Rasegård, 1994)

Chapter 8

GOAL-SETTING AND EXPECTED OUTCOMES

8.1. PURPOSE AND GOAL OF A SYSTEM

In an earlier context, the system concept had been discussed without the concepts of purpose and goal. However, it seems difficult to obtain a complete picture of a man-made system without mentioning both these concepts because they play a main role in a planning situation. Therefore, in connection to figure 16 these concepts will be discussed and put in the frame of the overlapping system concept.

Concerning the concept of purpose, Miller gives the following comments: "By the information input of its charter of genetic input or by changes in behavior brought by reward and punishments from its suprasystem, a system develops a preferential hierarchy of values that gives rise to decision rules which determine its preference for one steady-state value rather than another."(Miller, 1978, p. 39). Thus, in the words of Miller, a living system consequently is able to choose from a set of preferred values when it determines its purpose. A system itself is thus capable of deciding what shall be its purpose. Weinberg considers that " system has no purpose, for a purpose is a *relation* not a thing to 'have.' To the junk-dealers, General Motors exists to put out scrap-metal, yet the stockholders probably couldn't care less whether General Motors is producing cars or string beans, as long as it is producing profits" (Weinberg, 1975, p. 57).

However, there can be differences in the view of purpose depending on which viewpoint from within an organization an observation is made – from a corporate or from a subsidiary viewpoint. A subsidiary can see a purpose as its special purpose but this purpose together with similar purposes from the other subsidiaries form supporting purposes that are included in the organization's main purpose. So, what is the innermost meaning of purpose? To answer this question it seems possible to state that a purpose of a man-made system is in some way related to its main outputs as it is understood by the main part of observers, but that a system also can be related to various purposes at the same time, seen from the viewpoints of a lesser part of observers.

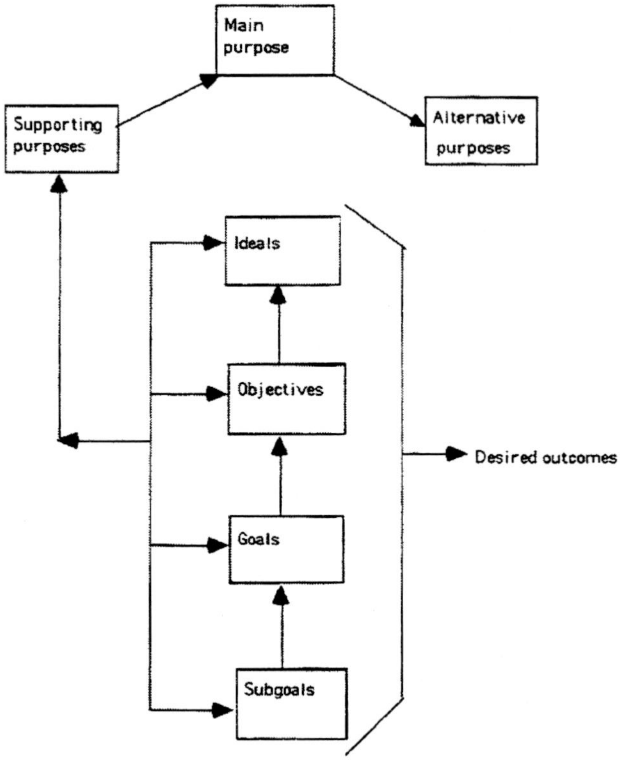

Figure 16. The purpose linked to desired outcomes (from Rasegård, 1998).

Concerning goals and objectives Kjell Samuelson gives the following definitions of these concepts:

> "(Ultimate) *objectives*: Are derived from and partly include desirables, have no other precedents; and are in common use termed 'objectives' being superordinate, precedents to goals. *Goals*: Have objectives as precedents, are often operational, can be decomposed into *sub-goals* and have several succedents as indicated by our structural model; goals can be updated in response to events in the external environment" (Samuelson, 1978, p. 200).

Professor R. L. Ackoff's views on the concepts in question are as follows:

1. *Goals*: Ends that are expected to be obtained within the period covered by the plan.
2. *Objectives*: Ends that are not expected to be obtained until after the period planned for, but towards which progress is expected during this period.
3. *Ideals*: Ends that are believed unattainable, but towards which continuous progress is thought to be possible and is expected. Therefore goals can be considered to be means to objectives. And objectives can be similarly considered with respect to ideals. (Ackoff, 1984, p. 41).

Miller's comments on goals of an organization can be summed up according to the following:

> " An organization has many goals simultaneously. They include general goals like putting a man on the moon, representing the majority of workers in an industry, or building a ward heeler into a nationally recognized politician, and specific goals like developing an effective heat shield for a spacecraft, organizing the workers in a particular plant, or getting a client elected to the state assembly. The long-term goals of organizations may conflict with their short-term goals. This can occur if a profit goal for a single year depends upon taking actions that diminish the probability of attainment of a long-term goal" (Miller, 1978, p. 651).

What has been said here about objectives and goals indicates that the desired outcomes have a hierarchical structure where each outcome is expected to achieve a desired outcome on a higher level. Certainly Miller is avoiding the term "objective" in his comments on desired outcomes. But instead of that he uses the terms "general goals" and "specific goals" as well as "long-term goals" with respective "short-term goals". Therefore it seems possible to translate his view of the goal concept into objectives and goals according to the definitions given above by Kjell Samuelson.

Concerning Ackoff's definitions of 'objectives' and 'goals' these definitions are not fully corresponding to what recently has been said because he argues that objectives can be obtained only after the planning period has come to an end. And that seems in some way impractical when one considers that the efforts ought to be directed at setting objectives for long-term planning, and measuring the degree of success in realizing these objectives during the planning period. Ackoff's concept of objectives seems impractical also for the reason that he uses the term 'ideals' for a desired ultimate outcome which is considered unattainable, but towards the progress thought to be possible and is expected. Instead, the meaning of 'ideal' ought to be modified to a desired outcome, which under ideal premises would be achieved during a long-term planning period, perhaps not before the end of the actual period. In this context it can also be of a certain interest to cite a couple of sentences made by one of Sweden's most prominent industrial leaders, Dr. Curt Nicolin. His statement can be translated into English as follows: "One fact is clear: The goals which do not have any point of time to be realized are not any goals at all. Furthermore, the goals must in any sense be measurable."(Nicolin, 1996, p. 49).

The conclusion of the reasoning about concepts of desired outcomes of a living system in general and an organization in particular, ought to be that it is possible to arrange the concepts in question according to the following:

1. *Ideals*: Towards which an organization is expected to progress even if it seems impossible to realize the desired outcome during a certain planning period, but which are not unattainable in a longer perspective, for example, a higher degree of improving the citizens general standard of living.
2. *Objectives or general goals*: Desired outcomes at higher levels which an organization is expected to realize during a certain period, for example, a certain degree of the traffic flow in a city.
3. *Goals*: Desired outcomes at lower levels which an organization is expected to realize during a certain planning period, for example, constructing a certain number of day-care places at day-care centers. Goals which ought to be formulated on an operational basis can consist of sub-goals.

8.2. SYSTEM ACTIVITY

> "A system is an object or process – or collection of objects or processes – that we choose to study. The environment comprises those processes that we expect to have an effect on the system. The inputs to the system are the effects of the environment on the system. The outputs of the system are the measurable responses of the system . The states of the system relate the inputs and outputs. The output function describes how the states determine the output" (Arbib, 1977, p. 176).

The comments on the statement above are delivered in connection to figure 16. The statement above describes a system as an object – or collection of processes that we choose to study. This means that a boundary of a system depends on what an observer desires to study. And the components within this boundary are those which have a special interest to the observer. The system's environment is also influencing the system through the inputs, and the systems refining of inputs in measurable outputs (goods or services). In the organizational system sketched in fig 21 the components of a living system with connected artifacts are indicated. The organization's inputs are marked by the "resource cake". These resources are used in the organization's activity (the state-transition function).

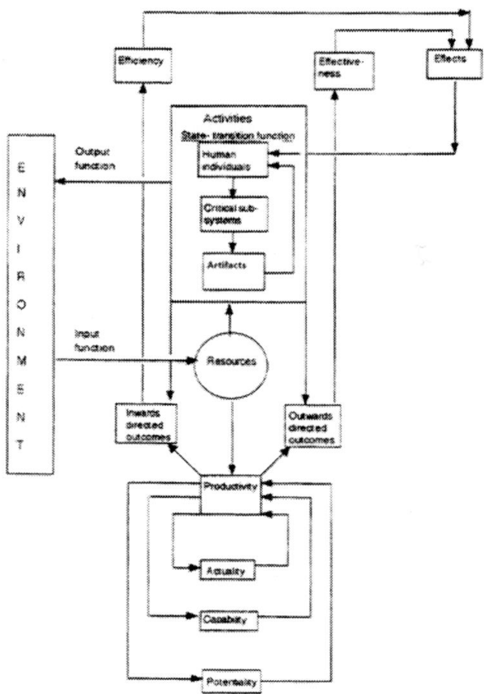

Figure 17. System related to an organization and its activities (from Rasegård, 1998).

From the activity the arrow marks the outputs that the system delivers to the environment which are determined by the states of the system depending on its activity.

However, the outputs must be related to some measures. Here Stafford Beer's three levels that measure a systems output have been used as examples of indicators of how the system's states change over time in response to inputs and the state-transition function (the organizational activity). The output function is related to the states of the system, in this example connected to Beer's measure levels. These are as follows:

"*Actuality*: This is simply what *we are* managing to do now, with existing resources under existing constraints.

Capability: This is simply what *we could* do (still right now) with existing resources, under existing constraints, if we really worked at it.

Potentiality: This is what *we ought* to be doing by developing our resources and removing constraints, although still operating within the bounds of what is already known as feasible (Beer, 1972, p. 207).

These three concepts from Beer can be directly related to productivity of different degrees. Thus, actuality is the lowest degree of productivity, capability is a higher degree, while potentiality is the highest degree. However, this highest degree cannot be reached before certain obstacles are removed and can often be a question of planning for a distant feature. As can be seen, the degree of efficiency as well as the degree of effectiveness, is related to the state-transition function and influences the human individuals in the leadership of the organization which can lead to a change of the organizational behavior. Apropos the meaning of efficiency and effectiveness may be, it can be reasoned, to refer to the following:

"*Efficiency*' and '*effectiveness*' are useful but vague terms which require clarification in terms of specific research goals. At times, unfortunately, these vague terms are only sufficiently clarified , but are used ambiguously. To avoid such equivocation, in the US Army research, we have elsewhere stated the following distinction:

Efficiency is viewed conventionally as '*doing things right*'; that is , efficiency involves a judgment concerning the internal workings of an organization.

Effectiveness is viewed conventionally as '*doing the right things*'; That is, effectiveness involves a judgment concerning how well the organization responds to its environment" (Ruscoue, 1982).

The definitions given above would perhaps make clear that an indicator is omitted, namely, productivity. However, here it can be remarked that productivity can be equal to efficiency and even to effectiveness. But it need not always be so. Take for example a building department within an agency, constructing a building for a day-care center for children. The subordinated sections of that organizational unit can excavate several cubic metres of earth and dig several metres length of tub ditches per unit time to a reasonable cost and, besides that, the work has been of good quality. Thus, the work mentioned shows a good productivity. But at last it will be clear that good productivity is not equal to efficiency. Set in a larger frame it will appear that the building is wrongly designed, so efficiency will be bad, even if productivity has been good.

So, the definitions of efficiency and effectiveness state that efficiency means *doing things right* while effectiveness means *doing the right things.* In the example of the day-care center the building was wrongly designed , and that means that the responsible building contractor had not *done things right.* But in spite of that, the section responsible had done a good job when excavating earth , digging tub ditches and so on. Consequently, seen from their viewpoint one can say that they have done things correctly. So, here it is possible to distinguish a dualism between efficiency and productivity. Therefore it seems reasonable to separate the concept of productivity from that of efficiency, and even from that of effectiveness.

An example has been given above of fulfilling a desired outcome, namely, to create a certain number of places for day-care children. That means that here it has been a question of creating conditions for a type of the agency's outward-directed social services to the citizens. This type of service has also to be related to some concept which indicates in what degree the given service corresponds to the desired outcome . And that is the concept of effectiveness. So, thinking once again of the day-care center it is possible to draw the inference that the fulfillment of a certain objective to give an increased service was bound to fail – the effectiveness of the service in question would be of a lesser degree.

Other concepts which can be of interest to look at in this context are the concept of cost-effective and cost-effectiveness. It is often said that production has been carried out in a cost-effective way, i.e, has been made in a cost-minimizing way in relation to the output gained. The output has in other words been produced in an economical way and considered the used resources. From that follows that the term cost-effectiveness is sometimes used to denote the special kind of effectiveness directly connected with the production of commodities and services in a cost-effective way. Thus, it would seem as if two more concepts have to be considered in addition to the three mentioned earlier. However, it seems unreasonable to use both the last-mentioned expressions as independent concepts for indicating a certain degree of realization of a desired outcome, because the concept of productivity can be said to involve both the adjective 'cost-effective' and the noun 'cost-effectiveness'. And therefore cost-effectiveness can be considered as synonymous with productivity.

As a result of the foregoing reasoning it seems desirable to use the following indicators of the degree of achievement of desired outcomes, in hierarchical order:

Productivity as an indicator of how well a producing unit has managed to use the inputs in relation to achieved outputs or, in other words, how well the spent resources of various kinds have contributed to perform the desired outcome in the production process.

Efficiency as an indicator of in what degree an inward-directed outcome has been fulfilled.

Effectiveness as an indicator of in what degree an outward-directed desired outcome has been fulfilled.

As can be seen in Figure 15 the degree of efficiency as well as the degree of effectiveness are united in effects. These effects are related to the state-transition function and influence the human individuals in the leadership of the organization and can lead to measures which influence the proceeding performances within the activity and is also outward directed and leads to a change of the organizational behavior. Thus, in figure 21 an attempt has been made to summarize indicators of desired outcomes with respect to Beer's three concepts, and assuming that the activity in question indicated in the figure is structured to the principles of the critical subsystems of Miller.

However, productivity must reflect the meaning of efficiency and effectiveness if the performance is to contribute to bringing about the desired outcome and related positive effects. In another case, a measured productivity can be denoted as a false productivity. It is also important that the planners in their planning point out which possibilities exist to realize certain objectives or goals at different levels of the organization in question.

8.3. Some Words about the Learning Function

Learning behavior by a system consists in its giving to the same class of inputs or stimuli a new class of outputs or responses , different from the class of responses given before. Learning responses may be primarily behavioral, consisting in directly observable changes in external output; or they may be primarily cognitive consisting of changes in the memories stored within the system and applied to the internal steering of its externally observable behavior; or cognitive and behavioral learning may be combined in various ways. Thus a system may acquire new patterns of information that enables it to perform reliable operations of discrimination or recognition, which it could not have performed before."

(Kochen, 1975, p. 62).

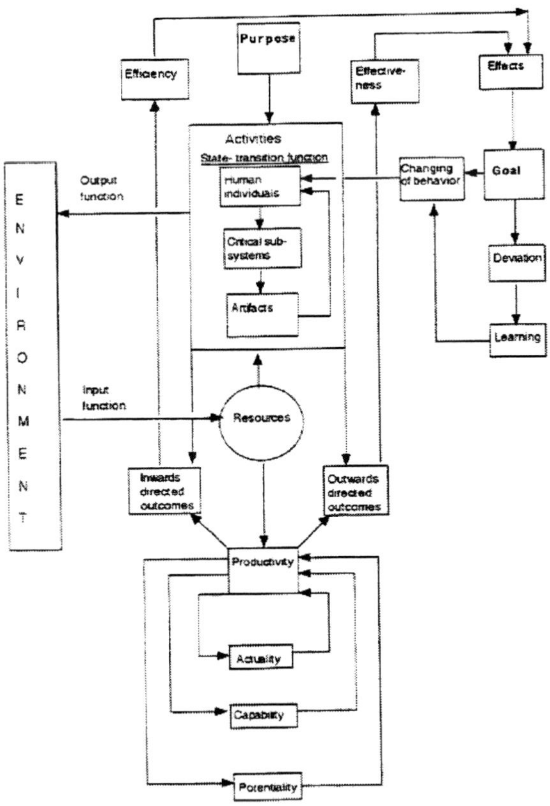

Figure 18. Purpose and goal related to a system of organization and the goal deviation´s influence on the learning function (from Rasegård, 1998).

Looking at fig. 18, which can be regarded as a completion of figure 117 commented earlier, we can recognize the features mentioned in the above citation from Kochen. We can see that deviation from the goal function influences the learning and can lead to a changing of behavior which in turn influences the state-transition function and also influences both the output and input function in connection to the measures of productivity. In this connection the question can arise whether the purpose of the organization is influenced through the deviation from the goal function. The answer seems to be that if the purpose would be put in question and be replaced with another purpose it will mean a changing of the organization from one type of organization to another , for example, if an organization's main purpose is to produce cars and it changes its purpose to producing refrigerators.

PART TWO
INFLUENCING SOURCES ON SCIENTIFIC RESEARCH

Chapter 9

SCIENCE AND RESEARCH

9.1. DRIVING FORCES AND TYPES OF RESEARCH

Which forces underlie the scientific research? To answer this we can have a look at figure 19 below.

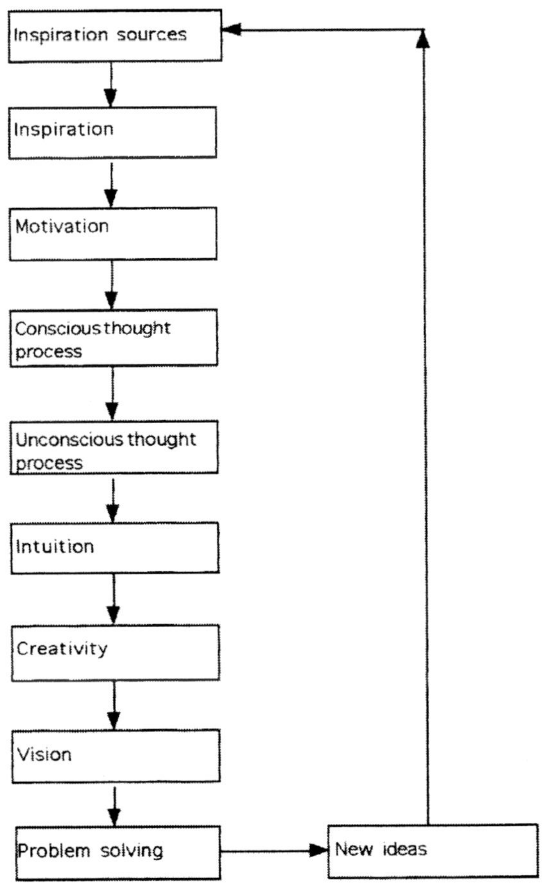

Figure 19. Some driving forces in scientific research (from Rasegård, 1999 b).

As can be seen we first have to seek some inspiration source depending on the researcher's mood and interests , for example, classic literature, music, arts, and publications of other researchers' experiences. Such an inspirational source can encourage the researcher to go on. Because the author of this report is conscious that the figure above gives a rather schematic picture of psychological elements involved in scientific research, it seems necessary to, in some way, clear up the meaning of the exemplified concepts. To begin with, the word "inspiration" comes from Latin "inspire" (breath in). The meaning of this word can be described as an impulsive thinking caused by an inspirational source. The inspiration has given birth to the motivation to go further with the research work. The motivation can, according to the Swedish Professor of Informatics and Systems Science, Kjell Samuelson, be defined as follows:

> "*Motivation* A complex phenomenon which refers to these things which cause us to move or act toward a goal. Motivation can be furnished by providing a situation where it is feasible to achieve certain physical, economic and social goals – although the aspects may interact to determine the resultant motivation. All these striving conditions are variously described as wishes, desires, needs, drives and the like. A motive is an inner state that energizes, activates, or moves (hence 'motivation'), and that directs or channels behavior toward goals. This result can be inferred from purposive, means-end behavior." Comparing "inspiration" and "motivation" it can seem as though these concepts would be identical. But it is not so. Inspiration can be like a switch in an electrical line, which is turning on the current from the inspiration source. Whereas motivation can be described as the flow of current which begins with the desire to solve a research problem. And that desire starts a conscious thought process. The researcher begins living with his problem – the brain is working on it. However, the researcher suddenly can have a feeling that a certain route is the right way to solve the problem. Despite that, he does not have a fully clear view of why the chosen way can lead to the goal , in other words, he has a fingertip-feeling or intuition as a result of the brain's subconscious work. Now his creativity will be tested. As in all production processes the research process involves problems which can be tackled in various ways depending on the creative ability of thinking by the researcher involved. Creativity, namely, in this connection can be described as a special ability of new-thinking and re-thinking of a human individual. The creativity originates a vision, that is, a whole picture of what the research problem involves and how it will influence the scientific solution. And now it is time for problem solving which, if successful, will create new ideas serving as an inspiration source for future research objects of the researcher in question as well as other researchers within the same field.

9.2. Basic and Applied Science

The basic research can sometimes be seen as conflicting with the so-called planned research. The latter type of research is often bound to applied research while the basic research mainly can be estimated as a more free research with not so strong ties to some form of social planning. In the following figure (Fig. 20) an attempt has been made to, in a simplified way, give an account of some main elements of scientific research and the connections between basic and applied research.

To begin we can see that applied science is influenced by basic research . Here it can be of certain interest to refer to the Australian Professor of Physics, Hanbury Brown, when he gives examples of the discoveries of penicillin as well as X-rays which had nothing to do with any form of planned research. These discoveries were made by accident (Brown, 1986). Another example is in physics with discoveries of the conditions of outer space which have contributed to develop applied science in the form of space technology that have enabled the development of satellites, space-shuttles and space-ships for exploration of the moon and other interesting astronomical objects. However, the basic research leads to forming new hypotheses, like Einstein's theory of relativity, which have to be verified. The science on this new ground leads to the fact that the discoveries founded on the verified hypothesis can break the ruling paradigm under whose umbrella scientific research will be done. These new discoveries can in turn cause problems which may then cause problems which have to be solved by scientific research.

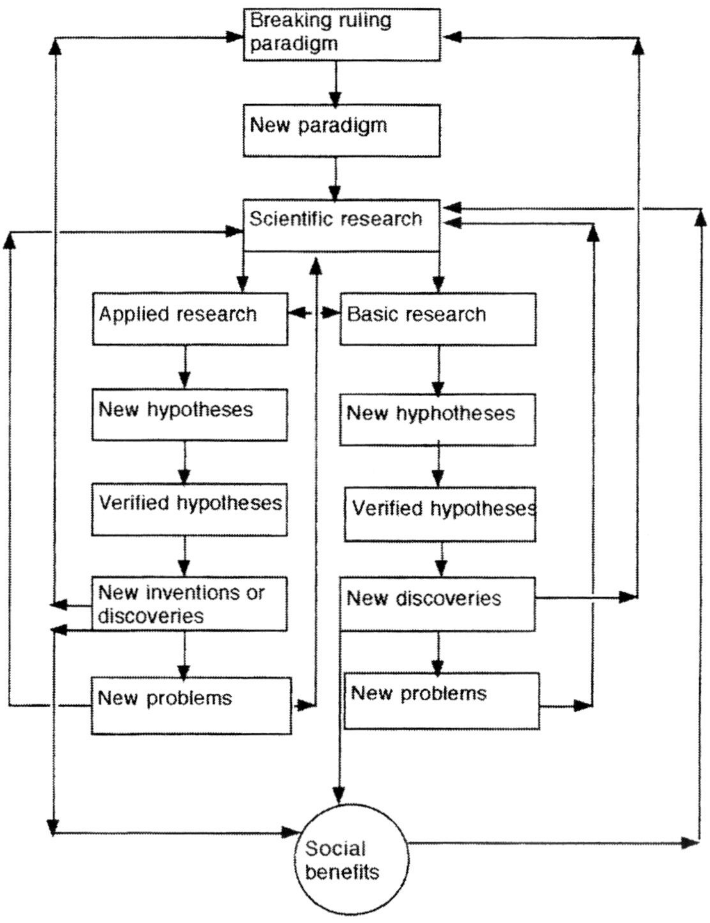

Figure 20. A simple structure of the scientific research process (from Rasegård, 1999 b).

If we continue and have a look on the left in the figure we can see that even the applied science leads to forming new hypotheses which have to be verified and can lead to new inventions which can confront a ruling paradigm. However, it can be noted that the verified hypotheses in the form of new inventions and discoveries which, when taken in use, can trigger a chain of events that can contribute to the arising of new problems , for example, the exhaust problems of motorcars and allergies caused by electrical devices which have to be solved by scientific research.

Concerning breaking paradigms when it comes to applied research we can take the social sciences as an example. Within this sphere it can be such that a research work within the branch of accounting and economy can lead to a new view of the economic and financial management system and in turn lead to a social invention which can break the existing paradigm of this branch of research. That research can, for example, begin with applying a new management system to an organization founded on theories of applied science and end with a result which either demands a revision of the existing paradigm or breaking it in order to form a new paradigm. At last both applied and basic research contribute to social benefits that can lead to creating a better world for mankind and creating new resources, intellectual and material, for further development of different branches of science.

So, it seems possible to argue that basic and applied research are two main types of research which can influence each other. The basic research is underpinning the applied research with new discoveries which have to be applied in different branches of the community of science. An example can be the aforementioned X-rays. On the contrary it can be said that even the applied research can influence the basic research. For example, through inventions produced as results of applied research which can be used by researchers of basic research, e.g., more advanced methods and instruments within the research of physics and medicine.

Thus, even within the social sciences it is possible to state that there are research-objects which can be classified as basic respective applied research. It can , for example, be a question of basic research which leads to a new economic theory which later on can influence applied science in the form of , for example, new management systems.

In this context it can be significant to again mention the Austrian Philosopher Karl Popper's ideas about objective respective subjective knowledge. The following citation can serve as an illustration of the core of his ideas: "It is interesting to note that most philosophers, though not all, discuss only knowledge in the subjective sense or (as I will say for brevity's sake) subjective knowledge. There exist many philosophical books devoted to the theory of knowledge, a theory which is also called 'epistemology' which never mention that there is such a thing as objective knowledge. And if they ever discuss objective knowledge, most of them assume that objective knowledge can be fully explained in terms of subjective knowledge, in other words, it is assumed that objective knowledge consists of many elements of subjective knowledge somehow linked together. I may tell you from the very start that for at least thirty-five years I have taken precisely the opposite view - without, however, making much impact. So here is a point where you will do well to look critically at what I am going to say. My position is this: I am mainly interested in objective knowledge and its growth, and I contend that we cannot understand the first thing about subjective knowledge and the give and take between the two kinds of knowledge (where subjective knowledge is more taking than giving" (Popper, 1994, pp.3-4). Popper gives examples of objective respective knowledge according to the following:

> “ (1) We may say: ‘It is well known that water consists of hydrogen and oxygen; or ‘it is well known that we can explain atomic and nuclear structures in terms of so-called elementary particles, but it is not known whether elementary particles have structure in their turn: ’This is still an open problem. These examples explain what I mean by ‘knowledge’ in the objective sense.
>
> (2) The following examples may explain knowledge in the subjective sense:
> ‘He knew he was exceeding the speed limit’.
> ‘He knew that water is composed by hydrogen and oxygen”.

If we take Popper’s ideas for granted we can go back and look at Figure 2. Then it seems possible to interpret his ideas in this way : The objective knowledge can be related to basic research while the subjective knowledge can be related to applied science (see the reciprocating influences in the figure).

Now it can be mentioned once again the necessity of which a society must afford to have a free basic research not tied up according to political trends and paradigmic obstacles. In this context it can be of interest to refer to the Russian Professor of Cybernetics and Mathematics V.V. Nalimov. He says among others: “I emphasize that a paradigm is, in my understanding, a stabilizing section, i.e., a protective mechanism, shielding at a certain stage of its development some trend from pollution or from spreading into lateral infertile areas. At another stage it may hamper the emergence of new trends. Every outstanding scholar begins his career fighting against some old paradigm and creating a new one, his own”. And he continues: “The statement of Norbert Wiener that 90 percent of the original work is done by 5 percent of the mathematicians, the remainder only playing a protecting role in sheltering it from being polluted by insufficiently strict constructions is widely known” (Nalimov, 1981, p. 10).

However it is not without risks for a scientist to make attempts to break an established paradigm which the following example from the U.S. shows: The example deals with how a young and promising pathologist at Harvard University who made the startling discovery that high blood levels of homocysteine could lead to arteriosclerotic plaques. This conclusion was startling because everyone knew that it was cholesterol that was responsible for arteriosclerosis. But the researcher in question knew that he was right so he published his findings in the American journal of Pathology putting forth a theory that something other than cholesterol caused heart disease. By 1979 his office had been moved to the basement, he was denied tenure at Harvard, he was refused funding for his research, and he was finally fired and was blacklisted. However, at last after several years his discovery was accepted and his reputation was restored and recently he was given a scientific prize (Greyson, Bruce, Journal of Scientific Exploration, vol. 15, pp. 3-4).

In relation to what has been said above it seems to be a question of great importance to give support to researchers of daring scientific ideas, even if these ideas can be in conflict with a certain ruling paradigm. Thus, a free basic research is necessary. This is a research not tied up to social planning and measures of quality, accountability and social relevance. A country which does not consider that she could afford free basic research risks stagnation and leaping into backwater. This section will be finished by coupling together Figure 19 with Figure 20 forming a new figure (Figure 21). This can be seen as an attempt to illustrate the individual scientist’s chain of thought elements from an inspiration source to problem solving within the framework of scientific research.

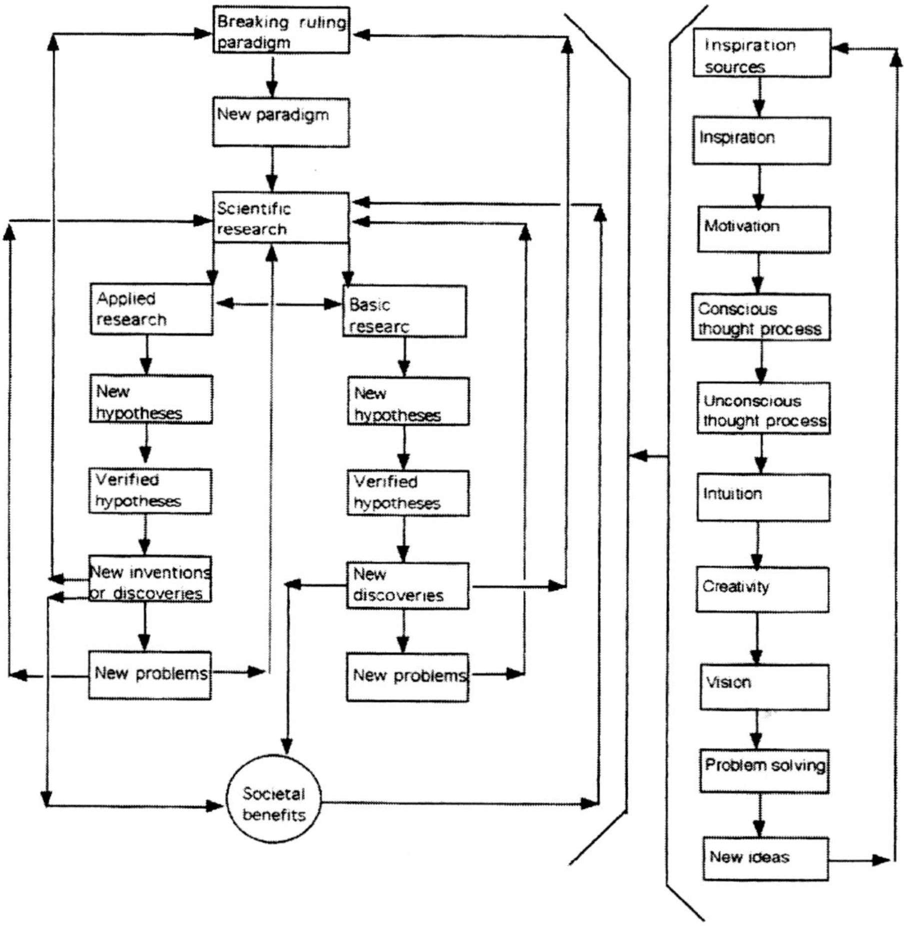

Figure 21. The individual thought process in relation to a simple structure of a scientific process (from Rasegård, 1999, b).

9.3. Dynamic Contra Completing Research

The British Professor of Physics, John Barrow, argues that some scientists devote too much of their costly time refining and smoothing results of scientific discoveries instead of working with research which is moving the frontier of science forward. He says that certainly the refining of the results of scientific research is a vital part of the scientific enterprise. But he is opposed to tendencies to "smooth away corners and rough surfaces" in order to shape artificial channels that lead from an imagined past ignorance and misconceptions to an enlightened present of "right thinking" (Barrow, 1991). Thus, a too exaggerated refining process can lead to a conformity and contribute to the fact that scientists with daring ideas, which cannot be put into this frame of conformity, can be stopped and in the best cases delayed in their research work. These calamities can be connected to influential researchers who are defenders of existing paradigms. Nalimov, also points out the problem of too much research which does not enrich science. He considers that many researchers, especially in mathematics statistics, produce what he calls "prestige publications". These publications, he considers, are of none or little value despite the fact that the authors make use of complicated and quite up-to-date mathematical methods. They are mainly written to enhance the author's prestige and are not of value when promoting future research leading to new discoveries (Nalimov, 1981). The questions connecting to the views above can be illustrated in the following figure (Figure 22).

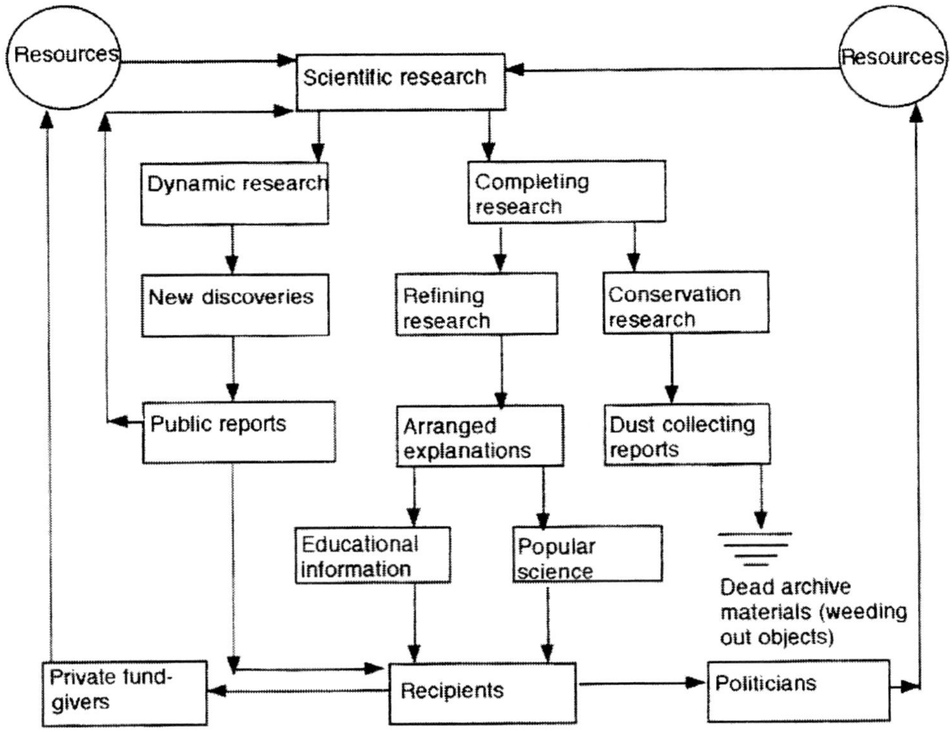

Figure 22. Different types of scientific research (from Rasegård, 1999 b).

As we can see from the boxes to the left in the figure, these boxes coupled to scientific research give an account for such types of research moving the frontier of science forward. We have here the type of research which can be denoted as dynamic research. This type of research leads to new discoveries which come to expressions in public reports (=publishing of facts). These reports stimulate other researchers within the community of science to continue with scientific research within the same field as the new discoveries represent. These public reports via mass-media also influence the recipients who consist of different categories of people in the society, for example, common people who elect the politicians who allocate financial resources to scientific research and private fund-givers who can be influenced by these reports, which means that new and increased resources can be available by these givers. As can be seen, the arrows from the recipients point partly to private fund-givers and partly to politicians. That means that recipients in this figure comprise individuals in private fund boards, politicians with the power of allocating grants, individual contributors and also students at universities and pupils in different schools (both basic and others) and last but not least, all other people in the society.

From the boxes to the right in the figure we can see that a part of scientific research can be denoted as completing research or, in other words, such research as Barrow mentions above. Following the left arrow from completing research we, as a consequence of this research, have a refining process with the aim to make certain results of scientific research more understandable . This work results in arranged information to be used in educational fields of various types, from basic schools to universities. The right arrow from arranged explanations to popular science indicates the efforts to bring the results of scientific research to the common people (recipients). As can be seen an arrow is also drawn from educational information to recipients. This indicates, as mentioned above, that recipients consist of different categories; students, politicians and working people in general, including people with influence on the private fund allocations, both donors, expected donors and members of boards of existing private funds for scientific research. The influence on common people is conveyed by mass-media channels, but also by open lectures and popular scientific journals and associations. But now over to the two boxes to the far right. There we see the box conservation research which is the type of research which Nalimov denotes as research connected to prestige publications . The work of producing these reports can be like a mill grinding empty shells instead of grain. That is a waste of the researchers' costly time for nothing. These reports are of little or no use and can in general be denoted as dust collecting reports in archives where they can be classified as dead archive materials which ought to be objects of weeding out. So, the meaning of the figure above is to show how the scientific research is chanelled and its influence on people of different categories in the society.

Chapter 10

PARADIGMS AND SYSTEMS INFLUENCING SCIENCE

10.1. NEW INFORMATION INFLUENCING SCIENCE

Apropos the forming of new paradigms and breaking down of old ones a simple schematic example from the ecological chemistry can be used to illustrate the meaning of the continuing process of scientific research.

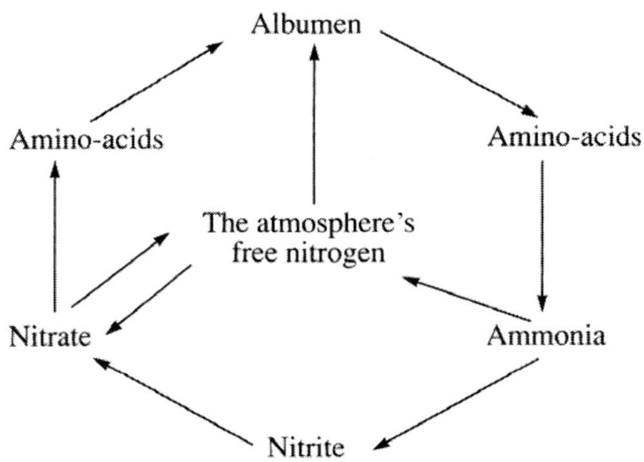

Figure 23. Circulation of nitrogen in Nature (from Rasegård, 1994).

The figure shows the circulation of nitrogen in nature in the process of breaking down and rebuilding albumen. Because nitrogen is an integral part of cellular protosplasm it is also a part of living creatures and plants. The albumen breaks down through influences by putrefactive bacteria into amino-acids that in turn break down into ammonia, at the same time it is united to the atmosphere's free nitrogen. The process continues and the ammonia is first broken down into nitrite which in turn is transformed into nitrate which is assimilated by the plant roots. This stage of the process takes an amount of nitrogen from and gives an amount of nitrogen back to the atmosphere's free nitrogen. The nitrate which is absorbed by the plants rebuilds amino-acids which in turn rebuild albumen - a process which also takes nitrogen from the atmosphere's free nitrogen. A phase of circulation is ended and the albumen is restored.

The chemical process in the simple form described above can be translated into the scientific paradigmic process according to figure 24 below.

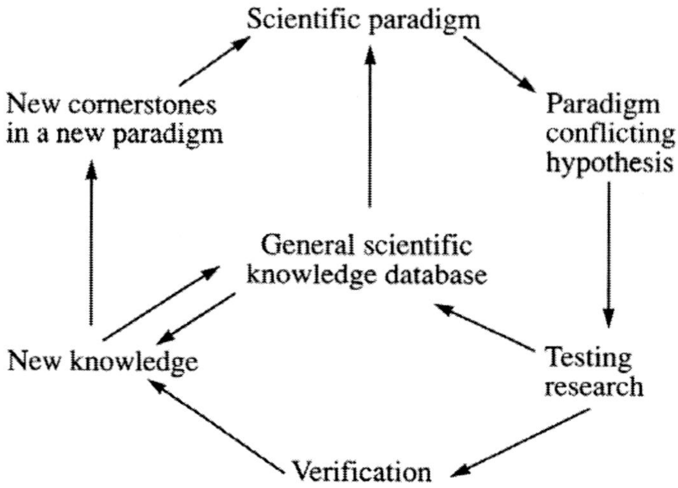

Figure 24. Breaking down old paradigm and rebuilding a new one (from Rasegård, 1994).

The break down process of an existing paradigm begins when conflicting hypotheses appear. These hypotheses are objects of testing research . This research results in the general scientific knowledge database receiving new materials which puts certain problems into focus which as spin-off effects can enrich the knowledge (the in-going arrow). The testing in turn results in verifications of the hypotheses. The verified hypotheses are published and this adds to the knowledge in the general scientific database (the out-going arrow) – at the same time as the new scientific theories are widely accepted (the in-going arrow). These accepted theories form cornerstones in a new paradigm . This new paradigm is continually influenced by the general scientific knowledge database (the out-going arrow). Thus, the figure above can be seen as an attempt to give a simple picture, analogous to the circulation of nitrogen in nature, of how a paradigm conflicting hypothesis after verification and acceptance can contribute in forming a new paradigm. In this context it can also be appropriate to mention researchers who are employed in what seems to be odd research sometimes lying out of the present scientific paradigm. Such researchers with odd ideas often have enriched societal development, even if their ideas at the time of presentation have been considered unfit for an existing paradigm, they gradually can be accepted.

10.2. Different Environments Influencing the Community of Science

In this section an attempt is made to elucidate some important environments which can exert influence on the community of science. When it comes to the different environments in which scientific research is done, however, it can stand to reason to continue on a while and look at certain features which are typical for the community of science. First, this community is not a homogeneous entity, rather it is a heterogeneous one split up like an archipelago of different islands. Certainly it can be stated that the authorities of universities and other scientific institutions as well as a superior authority authorized by the government are a uniting link, but that is only in administrative matters. When it comes to questions of pure scientific research fields, the splitting up is remarkable.

In Figure 25 below are exhibited different elements of networks which influence the community of science.

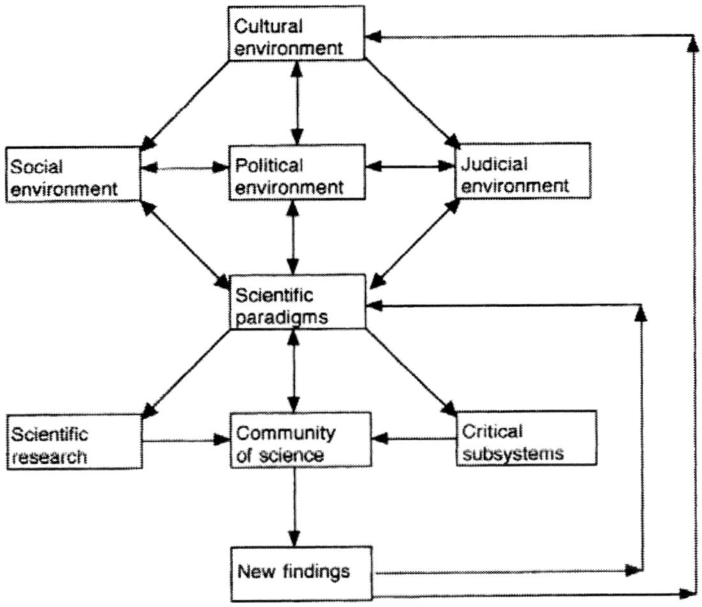

Figure 25. Some environmental sources influencing scientific research (from Rasegård, 1997).

As can be seen, the cultural environment forms the superstructure influencing the societal activities within the other environments. Concerning the political environment, it is interacting with the cultural. The politicians, as known, can influence the activities within the cultural environment, for example, by increasing or decreasing grants and other benefits. The norms within the cultural environment are also influenced by new findings within the scientific paradigms, for example, the educational area. When we study figure 12 we can also observe that the scientific paradigms are directly interacting with both the social, political and judicial environments and the question arises of how this is possible. The answer is delivered in the following.

The social environment is influencing the scientific research within the field of social policy and common welfare, which constitutes the paradigm of social science. The work within this paradigm also gives impulses back to the social environment which then can be conveyed further to the political environment in the form of proposals with the aim to improve the welfare system. The political environment can exert influence on all of the scientific paradigms through financial means (increasing or decreasing grants and other benefits). But these paradigms can also influence the political environment which can mean that the government issues laws and ordinances formed within the framework of a certain paradigm, for example, the ecological.

The judicial environment in the form of laws, ordinances and judicial experts influences the scientific paradigms, for example, by prohibiting certain medical experiments. But the judicial experts are also influenced by the thoughts ruling in certain scientific paradigms, for instance, the necessity to go on with certain biological experiments in order to combat serious diseases. Naturally, the scientific research within each branch of science is influenced by their paradigm respectively. The scientific research, of course, takes place within the world of science with its special demands of a pure science free from what is considered as any kind of speculative "science" or quasi-science. This world of science in this report has been named the Community of Science and is a part of the society even if it is connected to supranational associations. This community of science consists, as exemplified earlier, of different disciplines. So, here it is hardly possible to speak of any united association of science. Such an association is only formed for administrative purposes and consists, as mentioned earlier, of authorities connected to universities, other scientific institutions and governmental authorities to which the government has delegated its right of making decisions. Maybe, the scientific academies which are formed as kinds of forums for exchanging scientific findings can be seen as a kind of superstructure of represented branches of science, but any real power can hardly be ascribed to such associations.

The scientific paradigms also influence the individuals in the critical subsystems which are involved in the scientific process. The individuals connected to these subsystems are responsible for the research work within the respective discipline of science within the community of science. Thus, the box "scientific research" is aimed to point at the direction, purpose and goal of scientific objects of research while the box "critical subsystems" indicates the personnel involved in research and assisting functions. These individuals are influenced by the ruling thoughts within their respective paradigm. The research within different disciplines of the community of science gradually influences the respective paradigm. Thus, stone is laid upon stone in the building of an improved structure of knowledge and science as described in this attempt to give a picture of the many features of the relations between the society and the community of science. The cycle which is illustrated by Figure 25 is completed.

In the discussion of different types of information it can also be a question to insert two more concepts, namely, paradigmic and non-paradigmic information. The former is in accordance with a generally accepted paradigm. The latter is breaking the former paradigm. It is also possible to distinguish a third paradigm-connected information, namely, false paradigmic information. As an example of the last mentioned type of information the communistic paradigm-connected information in Eastern Europe during the time of the cold war can be mentioned.

If we stay on a while looking at the communistic paradigm it can remind us of a certain paradigm formed under the umbrella of this paradigm, namely, that of the Soviet geneticist Trofim Lysenko and his theory that acquired qualities could be passed on by heredity. His theory was, as known, accepted by Stalin and hereby received the official status as a part of the communistic dogma. The reason for the communistic dictator's acceptance can maybe have been that he speculated in the possibility that he, by indoctrination and brainwashing combined with biological selection, could form the Soviet citizen into an ideal Soviet-man, in other words, it seemed to be a possibility that, according to the communistic paradigm formed by the dictator, he could foster an inherited ant-hill mentality of subdued citizens in physically sound bodies.

Another theory of a similar kind was the theory of the Aryan race's superiority formed by Hitler. This theory, as known, was supported by certain "scientists". The theory was used as a device in order to motivate enslaving of non-Aryan people and to extinguish the Jews (the final solution). The denotation "untermenschen" was used in this connection. Both of these examples of two dictatorships' wrong ideas lead to the conclusion that we here are confronted with two main types of conflicting paradigms - the Western democracies' scientific paradigm and the dictatorships' quasi-scientific paradigms. The latter type was, as known, denied by scientists in the Western democracies.

In this context it is appropriate to cite Nalimov. He says: "Perhaps the most remarkable difference between science and all other fields is that in the process of its evolution, science *perpetually* makes its participants face new, more and more complicated problems which, in turn, allow a glimpse at the new mysteries of the World. The development of science is a growing dramaturgy of discovering new, previously unseen mysteries. Earlier, mankind faced new mysteries only with the advent of great teachers, such as Buddha or Christ, but now the discovery of new mysteries has become almost an everyday occurrence" (Nalimov, 1981).

10.3. Critical Voices about So-Called Established Paradigms

In connection with the foregoing section, it can be of interest to remember the American-founded organization Society for Scientific Exploration consisting of among others established scientists at well known universities, for example Stanford and Princeton in the USA and the Sorbonne in France. These scientists, besides their mainstream research, are employed in research in abnormal phenomena which cannot be explained within the present scientific paradigms. Here it can stand to reason to quote the editor of this organization's journal, Professor Henry Bauer: " Mainstream orthodoxy routinely resists novelties that later become accepted. Throughout the 20th century there are examples : Bretz's Spokane flood, McClintock's recognition of "jumping genes," Mitchell's insights into biological energy mechanisms, Woese's Archaea, and McCully's homocysteine. Only late in the 20th century did science reluctantly grant that acupuncture can have some analgesic effect, that ball lightning exists, that the kraken is not a myth but the real giant squid, that it is not foolish to look for intelligent life outside the Earth, that 5000-year-old megaliths incorporate substantial knowledge of astronomy, that human beings inhabited the Americas long before the days of Clovis' culture, and that living systems can sense not only electrical but also magnetic fields. Indeed, it may well be that the suppression of unorthodox views in science is on the increase rather than in decline. In *Prometheus Bound* (1994), John Ziman had outlined how science changed during the 20th century: traditionally (since perhaps the 17th century) a relatively disinterested knowledge-seeking activity, science progressively became handmaiden to industry and government, and its direction of research is increasingly influenced by vested interests and self-interested bureaucracies, including bureaucracies supposedly established to promote good science such as the National Academies, the National Science Foundation, and the National Institutes of Health. Parkinson's Law, as it may be, applies to science as well as to other human activities : no sooner has an organization become successfully established than it is considered already an obsolescent nuisance" (Journal of Scientific Exploration,

2000, number 3, pp. 304-305). In this context it also makes sense to point out the following citation: "Our knowledge is not in fragments. It is structured in the context of a paradigm within which we understand the pieces. Periodically, we need to restructure the context, hence we need a kind of system that tolerates the mavericks, the people who may be working on the germ of an idea that in no way fits the current paradigm. That may be the nucleus of an alternate way of organizing information when the flaws of the existing paradigm become more evident" (Kochen, 1975, pp. 25-26). Thus, it is obvious that a society must afford to tolerate certain persons with odd ideas not conformed to a certain paradigm, in other words, creative mavericks who maybe in the end can contribute to the breaking down of petrified paradigms for new inventions and discoveries. In this context we can remember Leonardo da Vinci who, as known, was much ahead of his time when he, for example , presented drawings of a submarine. In order to break down rigid paradigms it seems also necessary to seek a widened cooperation between scientific disciplines. Ultimately it is a question of utilization of new ideas for improving the conditions of mankind. Consequently, the scientific research must not be a mean to an end in itself – a kind of game which is devoted to "studying its own navel". Maybe, here can be reason to remember that a true scientist must have a humble attitude to his research field as formulated by a Professor of Astronomy (whose name I unfortunately have forgotten): "The Universe is an Ocean of secrets and we are only standing on the beach of this Ocean." Maybe, can the meaning of this citation also be applied to science research as a whole. In this connection it seems reasonable to quote the American Professor Robert G. Jahn of Princeton University. He says: "So, at the close of the 20th century, we find a monumentally extensive and complex scientific community that is more concerned with its applications, its economics, its politics, and its administration, than with advancement of its basic understanding. And regrettably, the priorities have been excessively reflected in our individual and collective public values, in our corporate and governmental initiatives, and in our educational strategies, at all levels. Thus at the dawn of the 21st century, we again find an elite, smugly contented scientific establishment, but now endowed with far more public authority and respect than that of the prior version. A veritable priesthood of high science controls major segments of public and private policy and expenditure for research, development, construction, production, education, and publication throughout the world, and enjoys a cultural trust and reverence that extends far beyond its true merit. It is an establishment that is largely consumed with refinements and deployments of mid the 20th century rather than creative advancement of fundamental understanding of the most profound and seminal aspects of its trade. Even more seriously, it is an establishment that persists in frenetically sweeping legitimate genres of new anomalous phenomena under is intellectual carpet, thereby denying its own well-documented heritage that anomalies are the most precious raw material from which future science is formed"(Robert G. Jahn, Journal of Scientific Exploration, 2001, vol. 15, pp. 23-24). Following from the same author concerning the actual opinions of leading scientists at the later period of 19th century, it can be cited: "A naive consensus abounded that most of the hard work of natural science had been done; that only mop-up tasks remained. As their towering patriarch Lord Kelvin (Thomson, circa 1884) proclaimed: 'There is nothing new to be discovered in physics now. All that remains is more and more precise measurement.' A sentiment echoed by their contemporary hero, A.A. Michelson (1894): 'The more fundamental laws and facts of physical science have all been discovered, and these are now so firmly established that the possibility of their ever being supplanted in consequence of new discoveries is exceedingly remote ...Our future discoveries must be looked for in the sixth place of decimals' " (Journal of Scientific Exploration, 2001, vol. 15, p. 22). Another citation from the same journal by Professor Henry Bauer, editor of the journal follows here:

One symptom of change, identifiable only in hindsight, was science´s failure from about the middle of the 20th century on, to satisfy public curiosity about mystery phenomena that arouse wide interest: psychic phenomena, UFOs, Lock Ness Monsters and Bigfoot. By contrast, a century earlier, prominent scientists had not hesitated to look into such mysteries as mediums, which had aroused great public interest. My claim here is not that UFOs or mediums belong in the corpus of science ; I am merely suggesting that when the public wants to know "What´s going on when people report UFOs?", the public deserves an informed response. It used to be taken for granted that the purpose of science was to seek the truth about all aspects of the natural world. That tradition had been served by the Mertonian norms: Science disinterestedly and with appropriate skepticism coupled with originality seeks universally valid knowledge as a public good.

These norms imply that that science is done by independent, self-motivated individuals. However, from the middle of the 20th century and in certain situations , some mainstream organizations of science were behaving not as voluntary associations of independent individuals but as bureaucracies. Popular dissatisfaction with some of the consequences stimulated "New Age" movements. In the 1980s, some scientists were led to form new organizations - notably the Society for Scientific Exploration and the International Society of Cryptozoology - specifically to pay attention to matters of public interest that mainstream organizations had been ignoring (Henry Bauer, Journal of Sciencific Exploration,volume 18, 2004, pp. 644-645).

From the same professor the following comments can be cited:

> "Minority views on technical issues are largely absent from the public arena. Increasingly corporate organization of science has led to *knowledge monopolies*, which, with the unwitting help of uncritical mass media effect a kind of censorship . Since corporate scientific organizations also control the funding of research , by denying funds for unorthodox work, they function *as research cartels* as well as knowledge monopolies. A related aspect of contemporary science is commercialization. Science is now altogether different from the traditional disinterested research, by self-motivated individuals, to understand the world. What national and international organizations publicly proclaim as scientific information is not safeguarded by the traditional process of peer review. Society needs new arrangements to ensure that public information about matters of science will be trustworthy.

Actions to curb the power of the monopolies and cartels can be conceived : Mandatory funding of contrarian research, mandatory presence of contrarian opinion on advisory panels, a Science Court to adjudicate technical controversies, ombudsman offices at a variety of organizations. Most sorely needed is vigorously investigative science journalism" (Henry Bauer, Journal of Scientific Exploration, vol. 18, No. 4, 2004, p. 643). In this connection it can also be of interest to cite the American Professor of electrochemistry John O´M. Bockris. He says among other things in the preface of his outstanding book *The New Paradigm*: "Some things described in the book will strain the reader´s acceptance. I ask only for an open mind. The truth about the present science is that it is simply inconsistent with many undisputed facts. So, what if some disputed facts are true too? Should we not give them a hearing, unbounded by what we were taught at the university?

Many "impossible" things seem to be turning out (experimentally) to be possible. The thing we need is a new paradigm to make the new results acceptable (The New Paradigm, 2004). Maybe, it can be reasonable to insert another citation from the same author. Nietzshe and Heidegger both thought that Now Knowledge came from Inspiration and Revelation and not from rational thought. Johan Gray thinks of the revolution in Western thinking based upon rationalism and science as having led to Cultural Disaster. In Gray´s view, it has given rise to the modern world, a civilization without values on its way to self destruction caused directly by resource exhastion (but indirectly because the Western public sees nothing against hedonism and that involves using up all the resources for one´s own pleasure now). The question is whether anything can be done about it, now that the negative consequences have been realized? The positive vista which Gray sees, is in the East, in Japan, Malaysia and perhaps in China. There the pre-Enlightment concepts exist, as they were never extinguished. No country is more modern than Japan, but most Japanese are basically Buddhists and share clear values. Japan is by no means only a materialistic society. The Enlightenment interrupted the earlier relationship of people with the Earth. Bacon introduced a cult of *measurement.* It destroyed traditions which had been in existence for millenia. It reduced spiritual consciousness for the majority of the people in the West. It diluted and pushed out the Greek philosophy which had come down to us through Islam and Aquinas" (Bockris, 2004, p. 38).

Thus, it is possible to state that the ultimate objective of all scientific research is not to avoid giving birth to conflicting hypotheses which can seem to break a ruling paradigm. Instead of that the aim must be to try out the conflicting theses in an unprejudiced way and widen the circle of knowledge in the human society in such a way that it can be used as a lever of lifting mankind to higher and higher levels of utilized knowledge and wisdom without spoiling ethical norms.

10.4. Different Main Paradigms Forming an Overlapping Paradigm

Earlier the paradigmic influences on the scientific research had been mentioned. However, common people and their acceptance or non-acceptance of events in society are also under paradigmic influences. Here it can be noted that paradigms from different fields of human activity together can form an overlapping generally accepted paradigm. This paradigm then is formed by the main paradigms of sciences, ethics and politics according to figure 26.

Consequently, Figure 26 can be seen as an attempt that in a simplified way point at certain main features in the forming of an overlapping paradigm. *The overlapping paradigm* (generally accepted) is then influenced by the ethical, political as well as the scientific paradigms. But which are the features of these paradigms? This question the author will try to answer according to the following: *The ethical paradigm* in our Western society can be considered to have its main sources in Christianity, foremost in the ten commandments and the Sermon on the Mount. These sources saturate our ethical feelings and within this pattern of ethics we are acting consciously or unconsciously based on the mentioned sources. *The scientific paradigm* as a whole protects recognized science from the influence of such research which is considered doubtful, for example, certain phenomena connected to spiritualism and the like or supposed visits by extra-terrestrial creatures, which phenomena can not be explained by our present physical laws (cp., section 4.). *The political paradigm* is connected to the ideology of democracy as the best system of governing a society with its different social functions. These three main paradigms form together the overlapping paradigm under which umbrella the common people are acting.

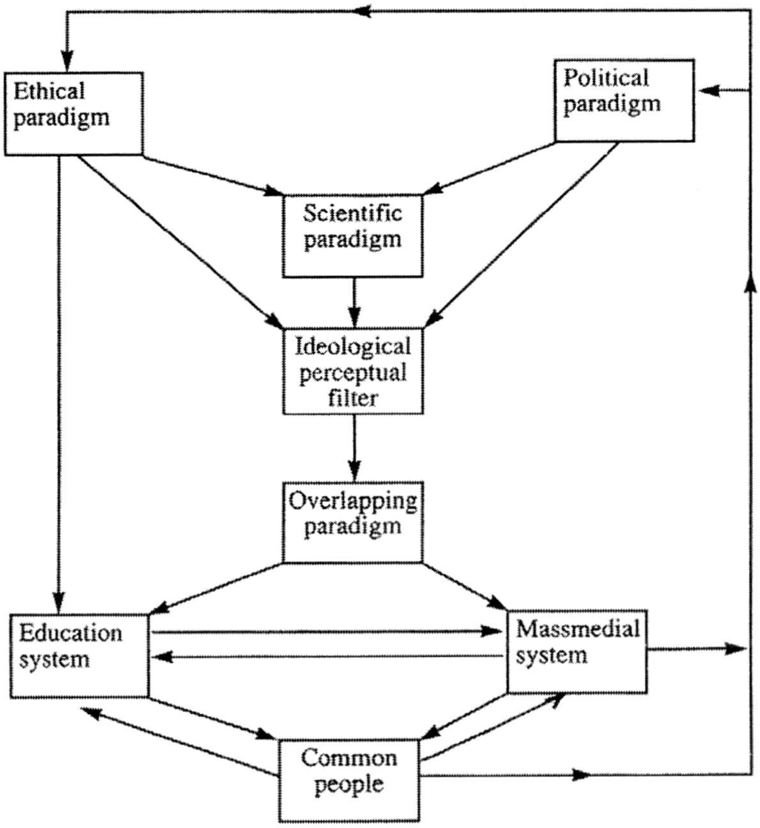

Figure 26. Paradigmic influences on societal opinions (from Rasegård, 1994).

As can be seen in the figure, both the ethical and the political paradigms are influencing the scientific paradigm. Within the ethical paradigm certain ethical rules for scientific research are formed, for example, prohibiting painful medical experiments on human beings and animals or gene manipulations in conflict with accepted ethics. The political paradigm influences the scientific paradigm in the way that politicians allow grants for scientific research and hereby can give priority to one branch of science in preference to others. From the three mentioned paradigms, the information passes through an ideological perceptual filter or, in other words, the information is more or less colored depending on the ruling ideas within the ethical political and scientific paradigms. After this filtering the information is incorporated as a part of the overlapping paradigm's generally accepted information. This information is further conveyed to the educational and mass-media systems both of which have an interactive influence on each other. The journalists are a product of the educational system. On the other hand, the mass-media system conveys information which within the educational system influences the student's image of the real world. Not surprising, both these systems have a determining influence on the common people. Furthermore, as can be seen in the figure, the ethical paradigm influences the educational system by its values of ethical norms. The mass-media system influences both the political and ethical paradigms by conveying opinions. The last mentioned system is also influenced by common people's opinions about societal everyday occurrences as well as overlapping matters. Furthermore, common people influence the educational system both directly by parent associations and indirectly by opinions conveyed by the mass-media system. However, it must be said that it is the common people who in democratic elections choose their representatives who ultimately are responsible for applying the main paradigmic ideas to the society. Certainly the leading party confesses its belonging to democratic ideals but only with the lips. In practice it has a position of something like a dictatorship. The legal opposition parties can be split up into different political program views and be too weak to challenge the party which has the majority in political assemblies. We will then be acquainted with a conditional democracy where the opposition parties live their lives overshadowed by the party in office.

Another danger is the formation of political "broilers" within political organizations where young men and women are formed for a political career. These people then have their given place in the political party apparatus which is then serving as a "broiler farm." In that way the society will be endowed with politicians with little or no experience of the practical world beyond their political sphere. This can be a serious threat to a healthy development of society. Further, the political decisions risk being conformed by a consensus mentality which can lead to mediocrity. In its eagerness to hold its power position the ruling party can be tempted to make populistic decisions often of mediocre quality which are also influenced by level-mindness . Therefore it seems possible to say that this mediocrity favors the appointment of mediocre officials in influential positions. Group thinking and pack behavior will tend to dominate the political arena supported by the "responsible" mass media which runs the errands of the leading circles and therefore often depicts themselves as "responsible and progressive" in their duties of defending the ruling party and its allied organizations. The journalists of the "loyal " mass media thus contribute in forming some form of one-eyed "objective" political perceptual filter. Here it can stand to reason to take a look at the following figure which is aimed to shed light on the mass-media influence when contributing in forming a political perceptual filter (cf figure 26).

The political perceptual filter creates a conformed political sense. The conditions of forming a conformed society have here been added with *"solidarity"* which in this connection is a misused concept because it is often used as camouflage for decisions with the aim of levelling the income for such people who have raised their income somewhat over what is accepted as an average income. These people have often earned their income from hard work and also in combination with long education and training -—especially scientists who often have big loans for studies. But they are not permitted to enjoy the fruit of their efforts. In the name of "*solidarity*" they will be deprived a good deal of the higher income through progressive taxes. All people shall namely show "*solidarity*" and give the *State* higher income by above average higher taxation as an aid to financing political decisions, often of doubtful nature with bad or mediocre results. The bad effects of these decisions then are silenced or belittled in the "responsible" mass-media. Instead a rather mediocre result will be applauded and described as progressive and successful in the "loyal" or responsible press. This is the "loyal" mass-media influence on the political perceptual filter.

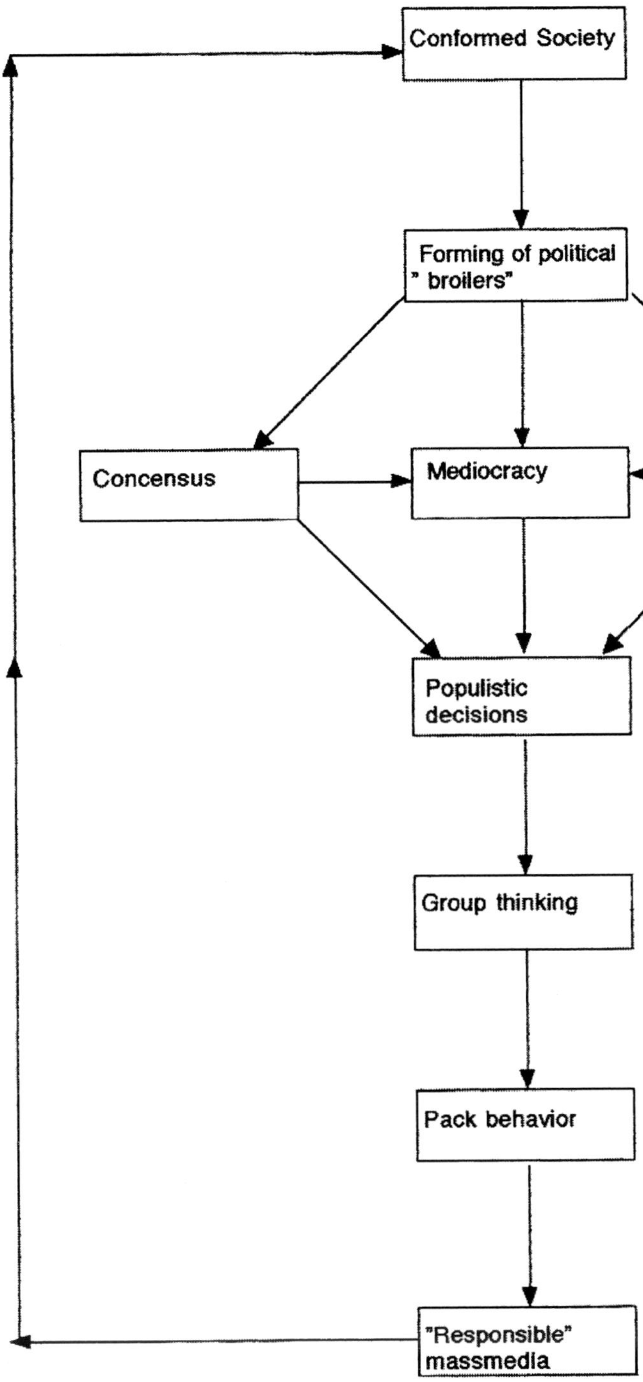

Figure 27. Certain components in forming a conformed society (from Rasegård, 1999 a).

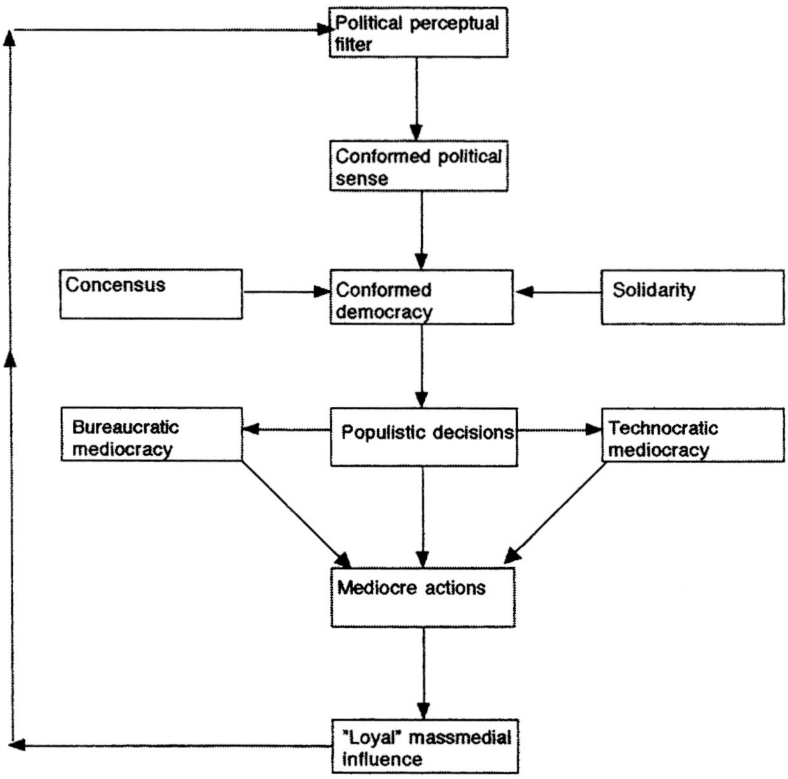

Figure 28. Mass-media influence on enforcing conformity (from Rasegård, 1996).

Consequently, the related tendencies in a political democracy reinforces the risk that society will end up as a type which the American author Ayin Rand describes in her social satire *Atlas Shrugged* . This satire deals with a corrupt and hypocritical society ruled by a dictatorship of mediocre bureaucrats and technocrats who are making preposterous interferences in the name of the people – thus creating the image that decisions were made in the name of a "peoples democracy" (Rand, 1957). However, as has been said earlier, ultimately the responsibility for acting to prevent these related tendencies is that of the citizens and comes to expression in the election of their representatives in different political organs. Perhaps it is appropriate to end with the following words:

> "In the German development, we saw millions of honest and decent citizens brainwashed to the extent of believing that the Nazis' rule was good for Germany and even good for humanity. Many people still believe in what they will barely admit is a totalitarian system, and will yet shudder from the name of Hitler and Nazism. A basic fact about human beings – even civilized human beings - is that their coating of civilization is often wafer-thin, but their ego is large. People believe they know what is best for others; democracy, as a result, is always in a tenuous and potentially unstable state (George, 1977, p. 67).

Chapter 11

Science on the Wrong Road

11.1. Non-Creative and Distilled Scientific Information

In the scientific work research reports sometimes can appear which, it is hard to say, do not give evidence of creativity , and despite this they can be denoted as scientific. Nalimov mentions these types of reports. He says: “In this context I emphasize that a paradigm is, in my understanding, a stabilizing section, i.e., a protective mechanism, shielding at a certain stage of its development some trend from pollution or from spreading into lateral infertile areas. At another stage it may hamper the emergence of new trends. Every outstanding scholar begins his career fighting against some old paradigm and creating a new one. His own”. And he continues: “The statement of Norbert Wiener that 90 percent of the original work is done by 5 percent of mathematicians, the remainder only playing a protective role in sheltering it from being polluted by insufficiently strict constructions, is widely known” (Nalimov, 1981, p. 10).

In an earlier context a statement by Nalimov about so-called prestige reports (cf section 3) was mentioned, therefore it can be worthwhile to again refer to a statement by Nalimov about mathematicians who within the ruling paradigm produce research reports and dissertation papers that lack creative imagination. Nalimov says: “In some countries mathematicians, especially in mathematical statistics, have begun to speak openly about the danger of so-called “prestige”, publications. These papers are written by many well-trained mathematicians who lack creative imagination but feel they must enhance their prestige . For this purpose problems are formulated with no concern for the utility of their solution or the reality of the initial premises. From the viewpoint of pure mathematics, these problems and their solutions prove of little value, though the authors make use of complicated and quite up-to-date mathematical methods” (Nalimov, 1981, pp. 159-160). So, here it is possible to state that it in certain cases it can be a danger for researchers to devote their costly time to produce “prestige” publications in order to show a certain scientific competence. How many dissertation papers, for example, of a type of “A study of the aphis erotic life” are not of use but instead are collecting dust in bookshelves hidden away in the libraries?

Maybe, it can be of interest that in the connection of what has been said above refers to some thoughts of man and science in a larger perspective concerning distillation of knowledge. Here it makes sense to refer to Barrow. He says: “When a new and deep idea is discovered for the first time, it may well appear in a cumbersome language principally designed to express some quite different set of ideas. Gradually, others will re-examine the discovery and find more succinct representations which relate it more naturally to existing ideas. This new relationship may be one of simple logical progressions from known ideas or it may be marked by a clash of opposites wherein one must make a radical choice between competing alternatives. This distillation of existing knowledge, to make it simpler and clearer, to refine the true metal of deep truth from the superficial dross surrounding it, is a continuous and vital part of the scientific enterprise. Some scientists are especially good at it and spend all their activities in furthering its aims rather than pushing forward the far frontiers of discovery.

We witness the consequences of this refinement process in many ways. It acts upon the history of science to distort its progress into artificial channels that lead from an imagined past of ignorance and misconceptions to an enlightened present of right-thinking. It smooths away rough corners and invents motivations. It places disparate individuals in an imagined community of like-minded researchers after truth. Yet, it does many of these deceptive things with the best intentions. *Is it not more efficient to teach a subject in a logical rather than a historical fashion.* The fact that past investigators took blind alleys is no reason to send others down them before revealing the better path. The result of this historical distillation is undoubtedly to make the laws of Nature appear increasingly simple, compelling, and altogether inevitable to our minds. In recent decades, the discovery of symmetry to be the master key which opens the secret door to the fundamental structures of Nature has been the prime mover in this never-ending quest for an ever more elementary image of things" (Barrow, 1991, pp. 114-115).

11.2. Smoothed, Commissioned and Biased Information

As a type of smoothed information it can be mentioned that in Sweden in autumn 1993 certain researchers in the Swedish Television stated that fat fish caught in the Baltic Sea contained too high doses of PCB and therefore was a threat to people's health – especially for pregnant women – and could be a cause of cancer. However, some days afterward the Swedish Board of Food Control stated that eating fat fish caught in the Baltic Sea twice a week is not dangerous. This can be seen as an attempt of smoothing the alarming statement from the researchers. However, these researchers could contribute to ruin the fishermen's outcomes as well as to related industries. Who is right? It is quite impossible for a layman to decide. Experts stand up against experts. Perhaps it is possible to say that this is a question of a sort of semi-truth. Yes, to eat fat Baltic-fish can be dangerous, but only if you pass a certain limit per week – a smoothed statement in order to calm down the people – both producer and consumer. Now the question arises: Presuming that eating fat Baltic-fish is a general danger to people's health, what is the value of healthy people as opposed to the loss of money in the fishing branch?

Even if it is true that a certain consumption of fat Baltic-fish is not a danger to health it can not go without saying that here is another problem revealed, namely, the scientist's integrity. In a conflict between science and powerful interests (economic or others) it is important that the scientists can work independently. This is so obvious that it can be denoted as a banality . As known, it is not usual that originally serious scientists can be trapped in a certain political paradigm or be indebted to powerful business interests. As an example of this commissioned information it can be mentioned certain tendencies in that type of research which can be denoted as social engineering where certain researchers can be inclined to satisfy the political establishment in order to get favors in the form of grants or other benefits. Another example can be taken from the pharmaceutical industry where certain scientists too willingly certify a medication's reliability. Such types of "scientists" who, as known, are not unusual in our boasted western democracy can be denoted as " prostituted boot-lickers". On the other hand it is not possible to overlook the fact that the temptation can be too strong for certain under-paid researchers with big study loans and heavy family obligations to devote their time to research of lesser quality in order to earn more money.

When conveying information to the common people, the journalists' biases play an important role, or to quote Nadler and Hibino: "Any information has a structured set of values and biases or inclinations. Eliminating them is difficult or impossible. What you can do is to be clear in identifying the values and biases that flavor the data. The journalist, for example, can be aware of his or her biases and strive to cover all aspects of the story, presenting it in value-neutral words. To write that someone "said" something, for example, has a far different connotation than to write that someone "admitted" something. To report that a cup is "half-full" has a decidely different tenor than to report that it is "half-empty" (Nadler and Hibino, 1989, pp. 207-208). It can also be a question to catch startling information for an organization or a society and conveying it in a right way without unnecessary smoothing or sensational adds (for example the Baltic-fish). In other words, it here can be a question of dealing with such kind of information as the British Professor Stafford Beer calls "algedonic" (Beer 1972) and which presumes an unbiased conveying without any kind of filtering.

Apropos this filtering of information it makes sense to refer to what Steven Rose says in the introduction to the book "Science and Beyond: "If in our methods of doing biology, the questions we ask about the non-human biological world, help determine which facts we collect and how we arrange them in our kaleidoscope, then *why* we ask the questions becomes as interesting as the questions themselves. We cannot observe the non-human biological world without bringing to it ideas about the organization of the world derived from our human experience. It is no accident therefore – and I do not mean this in any conspiratorial sense – that ideas developed in the interpretation of say, human monetarist economics, or the analysis of human sexual relations find themselves being called upon to explain the behavior of non-human populations from baboons to siamese fighting fish, mallard ducks and ants. We cannot avoid this cross-talk, we should at least be aware that it is happening and conscious of the consequences, which are what seems like 'pure science' is often pure science filtered through pure ideology" (Rose and Appignannesi, 1986, p. 7-8). This citation points out the fact that human beings are biased to interpret and draw conclusions from their own experience and in that manner form a perceptual filter which filters the perceptions, perhaps unconscious, in the same way as a journalist trapped in a paradigm, political or otherwise, conveys his/her perceptions to a news-audience.

In this context it can also be of interest to refer to the following citation: "Regarding objectivity, there is hardly such a thing, it does seldom exist. There are only different degrees of selling, from hard to soft; the latter being one handicap of the more modest and ethically concerned persons. When being your own salesman, you tend to criticise others and that is easy; but constructive criticism offering an improved workable and realizable solution is seldom presented. The critique tends to be directly proportionate to the critic's vested interest and rangership protectionism. The degree of criticism will depend on how bluntly the 'text strings' hit your preconceived worldviews , biases and emotions. It may also depend on what eyeglasses you were wearing (perhaps rose-coloured) and whether you had a bad day or are in a generally good mood." (Samuelson, 1978:1. pp. 329-330).

11.3. Misinformation and Wrongly Used Information

Typical examples of misinformation could be studied in the state apparatus of eastern Europe under the communist era. But even in the western democracies misinformation results from the ruling party's neglect, for example to give a complete picture of the real economic situation, or its overlooking certain unpleasant matters, not unusual in years of political elections. This use of misinformation can reinforce a rigid political paradigm and contribute to forming an ant-hill mentality of subdued individuals . Even in a democracy a special ant-hill threat of information nature is that the computerization of society will be so enlarged that human beings risk to lose their private integrity – everything has been on record, from the cradle to the grave. In other words, "big brother" sees and hears all about the citizens, a danger that has been rather threatening among others in Sweden. In this connection it seems necessary to point out the vulnerability of the human society which means that a computer net can be damaged or stored information can be used by individuals preparing a political coup. An example of that can be taken from the Kuwait crisis where the Iraqis, according to certain press information, tried to get hold of the computed national registration of the Kuwaiti citizens in order to manipulate a referendum which should express the citizens will to be a part of Iraq.

Even if the computers mean very much to the development of a society, a computerization can preserve old routines which in case of manual handling should be abolished and replaced by more rational routines. One can say that it then in some way would be like the robotic ant-hill routines. This can block as well a rational dissemination of information as preservation of old hierarchical structures in organizations and society. The former head of the Swedish board of supervision of the computed data registers, Jan Freese, has in the Swedish press stated that the computer technique can be the bureaucrats best friend. He among others says that the computer technique can be an excellent tool. But instead it has been the bureaucrats' most valuable asset when preserving a rigid hierarchical structure. Several times it has been a fact that what in other cases ought to be simplified administration which instead with the help of computers has been more complicated. Freese also coined the denotation "informations-atlet" in Swedish which could be translated into "Information Samson". These persons have a feeling that they only do whatever has been done. The difference in their opinion is that it now can be done by computers. They don't understand that changes are not always the same as progress. Freese continues with pointing out that rationalization for effectiveness in the public section must begin in the right end . It is the politicians who must start the process . In other cases it is a great risk that consultants of the "Information-Samson" type recommend computerization of old routines and regulations, and the client invests in expensive systems without a real effectiveness.

In this context it can be of interest to refer to some thoughts of Kenyon B. De Greene expressed in the journal "Behavior Science" volume 36, 1991. He points out that the major function of information technology has been regulating and controlling, and he uses the word "cybernication" to denote that. In the same article he also points out the danger of artificial intelligence because it can have a tendency to give too rigid or standardized information . In their eagerness to cover large areas of human brain activity the researchers can be trapped in too rigid paradigms. In both of these mentioned opinions from a man from the computer field (Jan Freese) and the scientist of neurobiology 90 (Kenyon De Greene) it is possible to trace a certain worry of rigidity of an ant-hill mentality.

Above the term cybernication has been mentioned as a denotation of cybernetic control of organizations and society. This is the negative side of cybernetics – that it can be used by dictators to tighten the control of society. However, the systems science has its roots in cybernetics. Then can this branch of science be used by oppressors for their special purposes? The answer is yes. But the systems science used in a right way contains in itself a seed of democracy because it means breaking up hierarchic structures and involves feedback through every level of an organization and a society.

11.4. Information's Weeding Out

> "I urge that we go deeper into elaborating and specifying what we mean with information. Problems of verification and validation of so-called 'facts' enter at this point. In general, as far as social science information is concerned. I feel that what we might be storing, if we store vast amounts of data, would, and well could, be myth and assumption rather than verifiable factual data. Indeed, I suspect we would be storing vast amounts of misinformation or non-information or out-and-out incorrect information" (Kochen, 1975, p. 173).

So, it seems most important to remind us that stored information must be scrutinized for relevance in the present time, especially international databases of a scientific character.

In this context it can also be of interest to refer to the well-known Professor of theoretical physics, David Bohm. He says in a book edited by Paavo Pylkänen: "so we can say there's not only information there's also misinformation. A virus in the DNA-molecule could be called a bit of misinformation, in that it enters the genetic structure and causes the cells to produce more viruses instead of more cells. In the same way as with cancer. Society too is full of this sort of misinformation. We have various ways of dealing with biological misinformation. The best way is by the immune system which recognizes it and gets rid of it, but we have no such system in society. Misinformation accumulates and society gradually decays. You see, the older the society gets, the more chance it has to accumulate all sorts of misinformation and the more it starts to fall apart. The society is blocked, because misinformation is held rigidly" (Pylkänen, 1989, pp. 80-81).

Miller says in this context: "Hypothesis 5.6-1 (see p. 110) (misinformation) can occur at the societal level. The hypothesis states that, if the system's negative feedback discontinues and is not restored by that system or by another on which it becomes parasitic or symbiotic, it decomposes into multiple components, and its suprasystems assume control of them. An example of a decider that stopped receiving negative feedback is Nazi Germany in its last days , when several decider components isolated themselves and received no feedback. In accordance with hypothesis 5.6-1, the system broke down, and the suprasystem – in this case, an international military government assumed control" (Miller, 1978, p.873). Thus, it is clear that it is not only a question of weeding out out-dated information, it is also extremely important that misinformation is not permitted to accumulate and that the functions on all societal levels receive an up-to-date continuing feedback of the relevant information in the society.

Chapter 12

A COMMON LANGUAGE OF SCIENTISTS

12.1. INTERNATIONAL ASSIMILATION OF SCIENTIFIC TERMS

In the international cooperation between scientists, it is a necessity that scientific and technical terms can be interpreted in a uniform way in order to be used in a right manner within the international community of science. In this context it can be of a certain interest to refer to the Japanese Professor of History and Philosophy, Shuntaro Ito. He states that when the Japanese scientists assimilated the terms of the Western science in their own language it was considered of utter importance to protect their cultural identity. Here follows his opinion in this question: "Indeed the very introduction of European science and learning into Japan came about through translation of technical terms. This was not common in Asia. In Indian, Islamic and South-Eastern countries there were no comparable efforts to use indigenous languages as channels for the acquisition of modern science. In Islamic countries, for example, higher science education is still generally conducted in English or French, while in Japan, of course, Japanese is used throughout.

The Arabic assimilation of Greek science in the 8th and 9th centuries shows a very interesting contrast. At that time, all scientific terms of Greek learning were translated into Arabic and embedded in Islamic culture. This would appear to be one reason why Arabic science made great original progress in the indigenous context in succeeding centuries. Japan's success in modernization, her lead over Asian countries and her continuing, it is suggested, are due to this initial Japanization of scientific terminology. This enabled her to avoid cultural colonization in spite of the large-scale adoption of Western learning, and to continue indigenous development thereafter.

Europeans, perhaps, overlook this unique point of cultural transfer sharing in the main, Indo-European languages, it is always possible for them to derive new technical terms from Greek or Latin for their common use. The Japanese case is different: Japan elected to adopt a European learning that was alien both linguistically and culturally, to her own. A similar situation can be observed in the modernization of China" (Shuntaro Ito in Traditional Thought and Ideological Shange, 1988, 9. 127).

However, even if the Japanese scientists and scientists of other nations translate technical and scientific terms from other languages into their own languages and invent new terms in their research works it requires, to be published in international scientific journals, that a language common to a wider audience, for example English or French, would be used. In their home countries respectively the researchers, of course, communicate scientific experiences in their own language. However, when it comes to publishing their results in international scientific journals and conferences it is necessary to use some language that covers a wider part of the scientific community. In figure 29 below an attempt has been made to, in a simple way, show some connections between language and research.

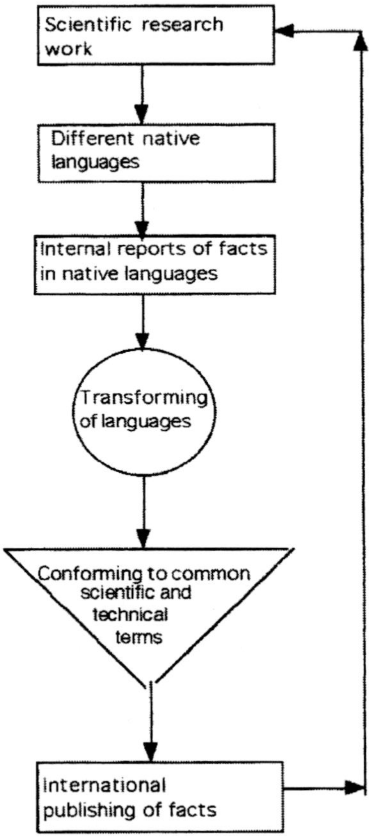

Figure 29. Influencing factors triggering events of the researchers′ accounting for results (from Rasegård, 1999b).

From the figure we can see that scientific research work is carried on in different native languages and results in internal reports of facts in the native language respectively. Furthermore, we can see a transforming of the native scientific vocabulary into another language which covers a larger language-area. That transformation leads to a conforming of scientific and technical terms in order to make them understandable to an international audience of researchers. The mentioned transformation makes it possible to publish the results of research in international scientific journals and/or make presentations at scientific conferences. And in this connection it is a question of putting the native scientific vocabulary into a common frame of science. This publishing and /or presentation can stimulate other researchers in their work that can lead to new discoveries and inventions which can enrich the community of science with new technical and scientific terms which then become common property.

12.2. DISSEMINATION OF FACTS

The above notes the importance of translating technical and scientific terms into the form which the scientists can use in their research work, and also when it comes to publish their results in the form of scientific terms. But what is a fact ? To the readers of this report the term fact is presumed to be clear. Nevertheless, it, perhaps, can be of a certain value to repeat the main meaning of this term. So, the publishing of facts means the end of a scientific experiment and investigation which has been judged in accordance with accepted scientific criteria . The term *fact* can be defined as a communicable human experience. When we say “that is a fact” we say so convinced that the person we address will be able to test our statement’s reliability. The person in question could, if he wants, exhibit the experience himself like we have done. Facts are communicable experiences in the way that they can be made common properties. Because the word “communicate” means “make common”, many scientific journals have in some form, communication in their titles.

When a researcher is publishing his result it is not, as many people think, a question only about reporting the factual circumstances. This individualized view neglects a most fundamental aspect of the scientific work – the sociological. The researcher reports his experiences and in the report is a careful description made of the environmental circumstances which influenced the experiences of his experiments or investigations work. These stated experiences can, according to the researcher’s opinion - usually with lesser varieties in the conditions - be repeated by other researchers in the same discipline .

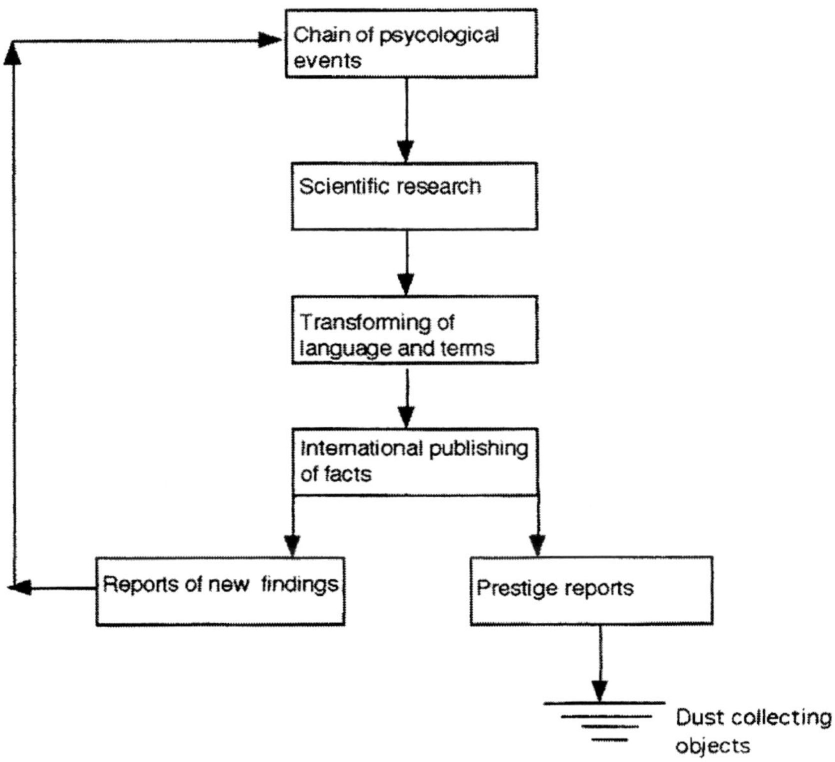

Figure 30. Main steps in a term-assimilation and publishing process.

If it is shown that the original researcher's observation can be repeated in other places it has been proved – not that he was right – but that his experiences could be made a common property and hence the individual experiences will be transformed into a fact. As a result of the researcher's published facts, new hypotheses can be formulated. These hypotheses can later on be verified and established as new scientific findings. These new findings lead to a moving forward of the science – but can also be breaking a ruling scientific paradigm which then becomes obsolete and replaced by a new one. An example of a changing paradigm in physics is the epoch-making discoveries made by Albert Einstein and Niels Bohr. At last we can have a look at the following figure (Figure 30).

This figure is meant to point out some main ingredients in a research and publishing process. Here a chain of psychological events (cf fig. 1, sect. 1) are starting the scientific research which ends with transforming indigenous language and terms, making it possible to publish facts for international audiences. These reports can then be divided into two types, reports of new findings and prestige reports. The former type stimulates the researcher's and other colleagues' chain of psychological events to engage in further research. The latter can leave other researchers indifferent and can be said to only affect archive personnel feedbacks as in Nazi Germany in its last days , when several decider components isolated themselves and received no feedback. In accordance with hypothesis 5.6-1, the system broke down, and the suprasystem – in this case, an international military government assumed control (Miller, 1978, p.873). Thus, it is clear that it is not only a question of weeding out out-dated information, it is also extremely important that misinformation is not permitted to accumulate and that the functions on all societal levels receive an up-to-date continuing feedback of the relevant information in the society.

Chapter 13

OVERLAPPING COOPERATION OF SCIENTISTS

13.1. NETWORKS OF RESEARCHERS FROM DIFFERENT DISCIPLINES

In this section it, perhaps, can be appropriate to point out the importance of Systems Science. Because the community of science is split up in many different disciplines and organizations it seems highly desirable to form some kind of interdisciplinary organ for observing research results and promoting cooperation between different branches of the community of science. In this role researchers of the inter –related discipline Systems Science have a big task, especially scientific editors of the journals connected to this inter-disciplinary branch of science. In order to point out the need for some interdisciplinary cooperation, the following press-cutting from the leading English-language newspaper of the Peoples Republic of China seems to be of a certain interest.

"Scholars Call for Research Coalition"

A number of Chinese scholars have appealed for the establishment of a coalition of researchers from the social and natural sciences in order to meet the increasing needs of the country's modernization drive. The present Chinese Academy of Social Sciences (Cass), a former division of the Chinese Academy of Sciences was formally established in 1977. Scholars said that the separation of the social sciences from the natural sciences in the country merits attention.

Scholars said that cooperation between researchers in the natural and social sciences is relevant to the present world which practices socialized production. This is true since democratic and legal construction requires not only a knowledge of the social sciences , but also knowledge of systematics, cybernetics, information and operational research.

They noted that improving the development of social sciences also requires the methods of contemporary natural sciences, modern mathematics, computer technology and systems engineering.

The scholars contended that the current separation of social and natural science researchers in China blocks the coordinated development of the fields.

They pointed out that a coalition of the social and natural sciences does not mean an institutional alliance. Instead, it should be considered as mutually beneficial for a future establishment of a new interdisciplinary branch of learning.(Xinhua)" (China Daily, February 8, 1991).

In this context it can be worthwhile to refer to a statement by James G. Miller. He says: "The spirits and the metaphors of our time account for the fact that many similar intellectual activities have been going on independently, one often quite isolated from another. Each worker or group has known something about a particular field, a chosen island, but frequently has been aware of related work in other disciplines . Oftentimes the same idea seems to have been discovered over and over again under diverse names in different areas. My efforts at integration have been partly attempts to establish a communication network among the islands" (Miller, 1978, p. XV). In connection to what has been said here about the desirability of forming some kind of interdisciplinary discipline which is overlapping other branches of science, a kind of meta-discipline, it can be of value to point out the following figure (Fig. 31). There the systems, cybernetics and information science form an emerging breed.

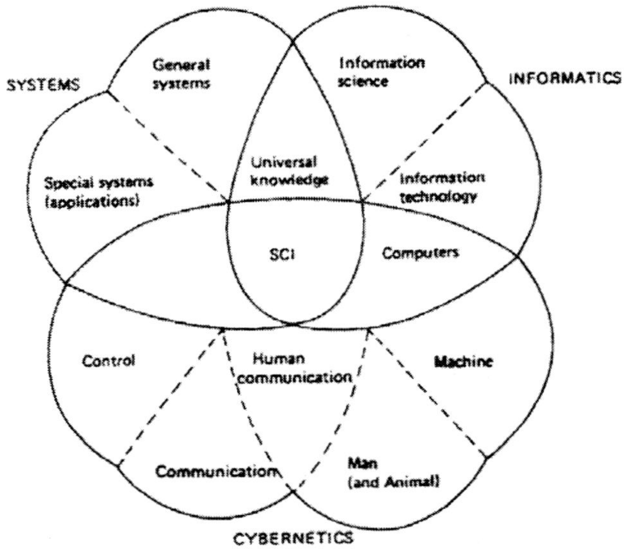

Systems, cybernetics and informatics as an emerging breed, to master the information systems' dynamics occuring at the fields' intersection (From Informatics by General Systems and Cybernetics, Samuelson, 1978).

Figure 31. Systems, cybernetics and informatics as an emerging breed, to master the information system´s dynamics occurring at the field´s intersection (from Samuelson, 1978).

Along with the figure above it can be of interest to refer to the following statement by Kjell Samuelson: "Disciplines are categories that facilitate the context of science; they are nothing more than filing categories. Dependence between sciences was taken for flow in only one direction. This hierarchical myth is still widely accepted in and out of science that is as severe and irrational as any that has existed in society" (Samuelson, 1978:2, p. 162). As examples of scientific disciplines where a cooperation is desirable it can stand to reason to refer to the following: "As far as we know, the human mind-brain entity is the most complex system of evolution. So, an interdisciplinary research program of computional neuroscience, physics, engineering, molecular biology, medicine and epistemology is necessary to deal with this complex organ. This is the reason why some scientists have started an interdisciplinary research program on the brain and mind with ethical and anthropological dimensions, which we have called 'neurobionics.'

In general 'bionics' means simulation of natural functions and processes by technical and artificial procedures and systems. The design of aeroplanes and submarines to emulate the aerodynamic bodies of birds and fish are well-known examples. Historically, bionics is an old dream of mankind to simulate the principles of nature by technical means to manage the complex problems of life. In this tradition, neurobionics means the elaboration of a common techno-biological realization of a brain-like computing system on silicon and/or organic materials for enhancing morphological and functional properties of natural neurons and developing neural prostheses. This is not the horrible vision of Frankenstein's monster. In order to become motivated and engaged in these goals of research, one needs only look at the desolate states of patients suffering from tumors or accidental injuries of the brain" (Mainzer, Klaus 1994).

13.2. Forming an Interdisciplinary Metadiscipline

As a consequence of what has been said above, the following figure (Fig. 32) exemplifies a cooperation of scientific disciplines in forming an interdisciplinary meta discipline.

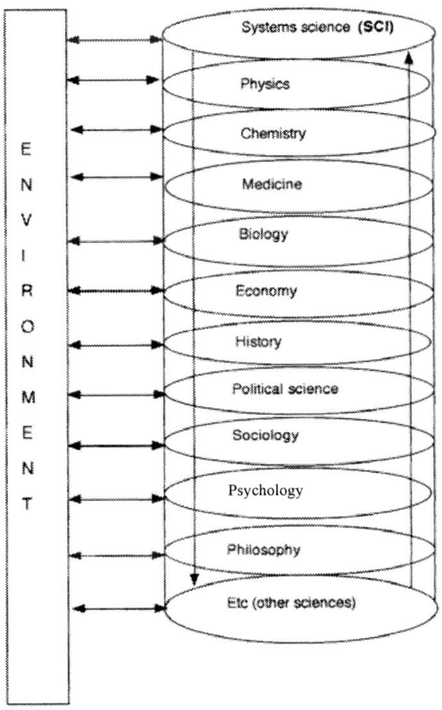

Figure 32. Example of slicing the community of science with the systems science as a source of cooperation (from Rasegård, 1997).

This figure is influenced by the Swedish interdisciplinary *informatics and systems science* of which the core is shown in Fig. 31. The contents of this discipline is mainly corresponding to what in international connections is denoted as *systems science*. Even if the author of this article is conscious of the fact that the figure only shows a fractional part of scientific disciplines the meaning is to give a hint at the necessity to form some kind of network between different disciplines. In this coordinating task, interdisciplinary disciplines like systems science have a great role to play. However, the question can arise of why such important branches of science as mathematics and linguistic science have been represented in the figure. The answer is that mathematics and linguistic science despite their importance in this context can be regarded as auxiliary branches of science serving as tools in the process of carrying on the branches of science exemplified in the figure. As can be seen in the figure, the exemplified disciplines have interactive connections with the environment which the double arrows indicate.

Now it seems reasonable to refer to the following: "Thus it is clear that the agenda of systems science is to fit a metasystemic framework over all scientific disciplines in an attempt to find interdisciplinary (as well as transdisciplinary) truths and relationships that will apply to each discipline in particular, as well as to all of them in general. To be such an all-encompassing discipline, systems science must be a *metadiscipline* , whose level of abstraction is above those of the other disciplines. Therefore, systems science is a general discipline whose body of knowledge lies at a level of abstraction similar to that of mathematics and philosophy". (van Gigch, John, 1991, p.236).

The rapid development in our time has, as mentioned, contributed to the creation of more and more specialized branches of science. In other words, the science shows a tendency to be divided into narrow strips where scientists such work without a great deal of knowledge of what is happening within other strips. The scientists may hereby become so specialized that their ability to communicate with other scientists outside of their own special field of research will be impaired . Thus, the highly specialized branches of science would contribute to form a scientist society which can foster a mentality of "branch-idiotics". Thus, the need for some type of generalists is brought into focus by the cited press-cutting – generalists who can be of help when building bridges between the scientific vocabularies.

Going back to the press-cutting from the Chinese newspaper it seems possible to divide democracy into two classes – one first order and one second order (to do that may be like a blasphemy or like "swearing in the church"). The first order is the characterized by a certain freedom within the economic sphere but a conformity in the political sphere. The second order is characterized by freedom in both the economic and political spheres. This is the form of democracy that what we call the western countries are confessed to – at least with lips.

Chapter 14

CERTAIN BASIC ASPECTS OF BEHAVIOR

14.1. SOME THOUGHTS OF KARL POPPER

Previously some ideas of Karl Popper had been mentioned. Therefore it will not be out of place to give a description of the core of his ideas of three worlds (Popper and Eccles, 1981). The contents and related concepts are briefly referred according to the following:

World III
The products of the human mind
6. Works of Art and Science (including Technology)
5. Human language

World II
The world of subjective experiences
4. Consciousness of Self and Death
3. Sentinence (Animal Consciouness)

World I
The world of physical objects
2. Living organisms
1. The Heavier Elements; Liquids and Crystals
0. Hydrogen and Helium

Karl Popper has presented these three worlds in a table (Popper and Eccles, p. 16) which he has named "SOME COSMIC EVOLUTIONARY STAGES" with the contents shown above. As can be seen the evolutionary development of our planet is divided into different stages from the creation (Big Bang) to the development of advanced products of the human mind. The following figure (Figure 33) can be regarded as an attempt to give an account for the expressions of human behavior related to Popper's three worlds.

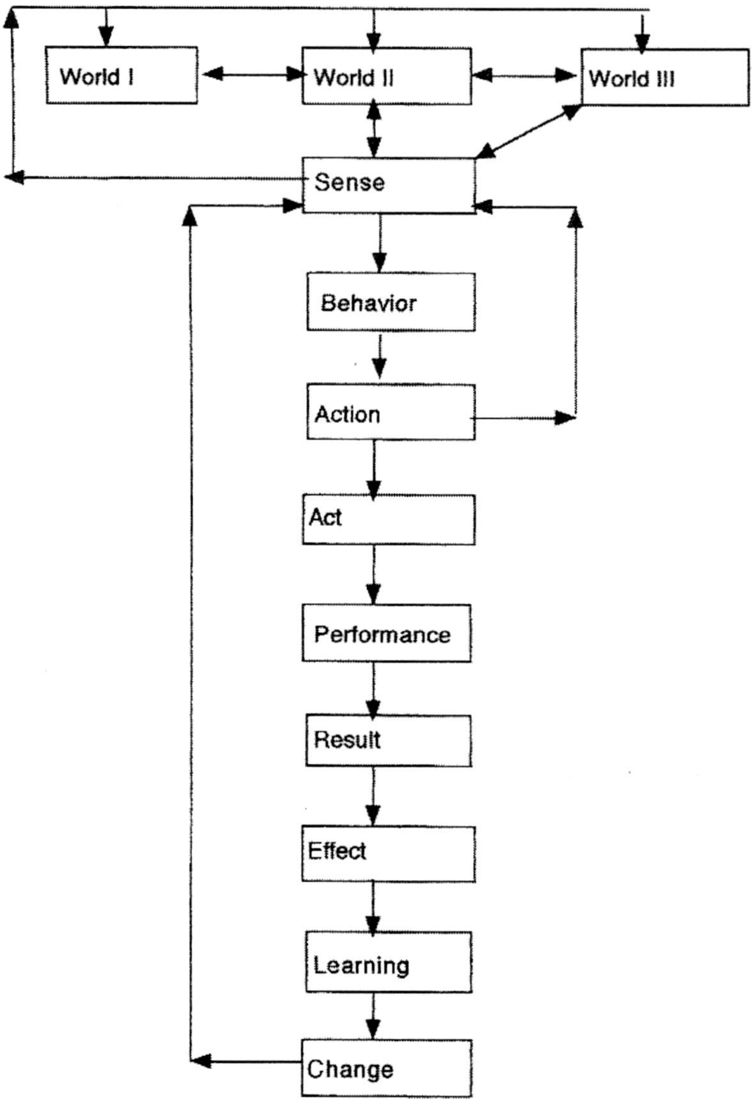

Figure 33. Expressions of human actings related to Popper´s three worlds (from Rasegård, 1996).

The figure shows that the three worlds have interactive influences on each other. World I influences World II , for example, by the ecological effects as results of human behavior when exploiting natural resources and pollution of air and water. The consciousness of World II is forming thoughts of the best way to preserve World I and to minimize the danger of destruction. These thoughts influence the human acting in World I. The consciousness of World II creates the prerequisites for development of World III and the products of the human mind in the form of cultural and scientific activities which in turn influence the consciousness in World II.

In the next figure (figure 34) the connections of environmental influences related to Popper's three worlds are shown. All three worlds exert influence on the cultural environment.

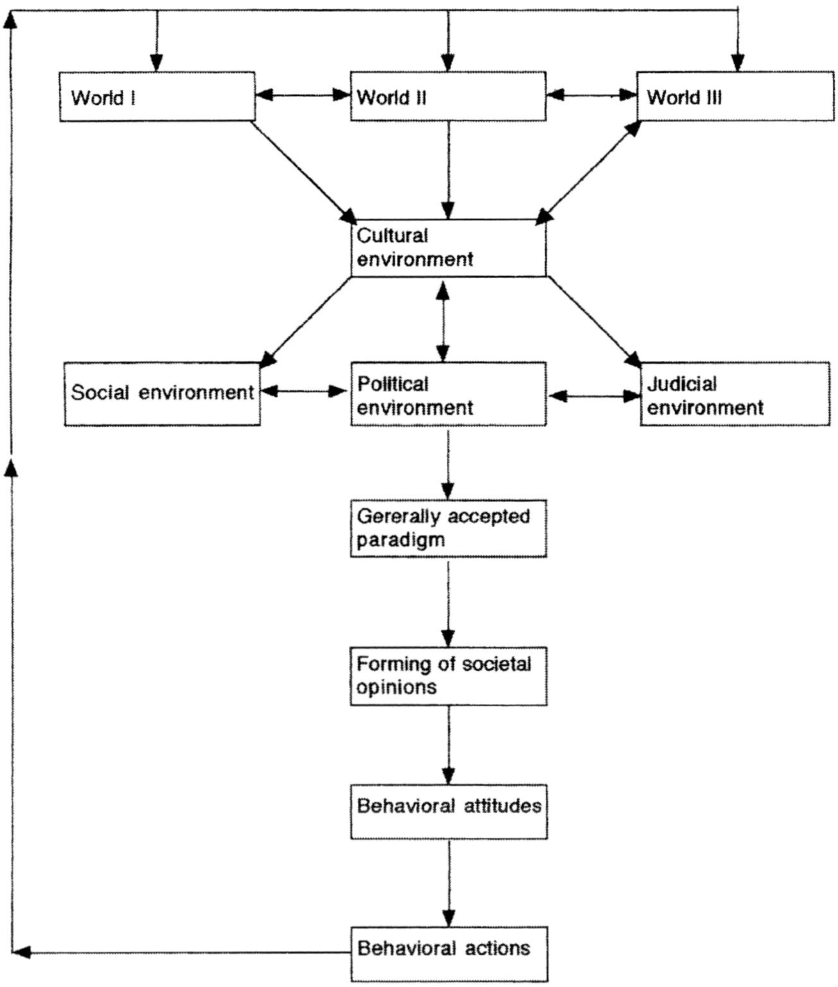

Figure 34. Human acting related to the cultural environment and Popper´s three worlds (from Rasegård, 1996).

Certainly one can deny that World I can hardly have any influence on the cultural environment . Nevertheless, the culture is a product of conditions of World I. The topography and climate, and other elements of World I are important when forming the external conditions of special cultures in different regions of our planet, for example, nomad culture in the arctic regions and culture based on farming in agriculture regions with a temperate climate. The consciousness of World II is influencing the cultural environment in a way that it contributes in the formation of the human self-consciousness which is a necessary attribute when forming a fundamental aspect of the cultural environment. World III is the most important source of influencing the cultural environment. However, it also receives stimulating impulses back from this environment. Thus, it is here a question of reciprocity.

In connection with the above it can be worthwhile to give a further citation from Karl Popper about these three worlds: "We live in a world of physical bodies, and we ourselves have physical bodies. When I speak to you , however, I am addressing myself not to your bodies but to your minds. So in addition to the first world, the world of physical bodies and their physical and psychological states, which I call "world I," there seems to exist a second world, the world of mental states, which I call "world II" . And so a question arises concerning the relationship between these two worlds, the World I of physical states or processes and the World II of mental states and processes. This question is the body-mind problem" (Popper, 1994, pp. 4-5).

As mentioned, Popper's three worlds influence the cultural environment. This environment can be regarded as a superstructure of the human society. Here the ethical norms ruling in the respective cultural area play an important , not to say dominating, role. The roots of these ethical norms are, as mentioned in an earlier context, mainly to be sought in religions and philosophical doctrines. The cultural environment which also includes scientific activities is influencing the social, political and judicial environments. The social environment is based on conventions of a social welfare society where the citizens have the human right of social care free from destitution. The political environment's main source of inspiration in the Western democracies is the conviction that democracy is the best way to rule a country. The judicial environment can be seen as an environment which is strongly influencing the co-existence of human individuals and organizations within a society. A main source of influence within these environments is the ethics formed within the cultural environment. But the most important influencing ideas come from the political environment through the laws formed by politicians. As can be seen in the figure the political environment gives stimulating impulses both to the social environment and the judicial environment and receives stimulating impulses back from those environments. It can also be observed that the political environment, aside from receiving impulses from the cultural environment gives signals back to the cultural environment concerning politicians' decided cultural and scientific activities following a certain political trend. The politicians, as known, can use their power to increase or decrease grants to scientific and cultural activities. The political paradigm contributes to form a generally accepted paradigm and to form societal opinions and behavioral attitudes that influences behavioral actions (cf. what is said later on in connection to the welfare system).

14.2. Certain Central Parts of the Concept of Behavior

In this report of man's relations to science and his environment, it is hardly possible to avoid devoting some thought to the concept of behavior as the steering factor of relations between human beings. Therefore, it can be worthwhile to give some examples of the contents of the concept in question, examples which the author of this report considers to be most important. As known, the concept of behavior is multifarious. Therefore, it is necessary to give an account for some definitions of the core of this concept. To begin with, the following simple fundamental definition can be given: "*Behavior:* The activity of an animal (human or otherwise) which an observer takes to be the result of genetic endowment and social conditioning" (Checkland, 1991, p. 312). A more detailed definition of the concept in question is given according to following: " *Behavior:* A high level abstraction which refers to an extreme large variety of actions which differ in form, duration, and frequency. We must identify and define a unit of behavior which can be pointed to and directly observed. We have chosen to use an *act* – a particular bit of behavior occurring at a particular time and place – as the smallest particle of society which is capable of being observed" (Samuelson, 1978:1, p.7). From the same source the following citation is taken: "*Act:* to actuate; to produce an effect or influence, especially a change in the conditions of something. A decision without follow-through actions is a block to achieving potential study payoff. Of course, there are decisions that are made to take no action in a given situation. In situations of logic, this is simply a recognition that at least for the present there is no net payoff for the actions now possible. The actions may be considered, but rejected. There are two subheadings under the verb "act". *Do* emphasizes performance from the point of view of the individuals involved and *change* stresses the result of the do actions" (Samuelson, 1978:1, p 2).

The British researchers L. Johnson and C.M. Tomlinson have given the following definitions connected to behavior: "In the social science a distinction is drawn between behavior, action and acts. The main terms at stake are "Verhalten" (behavior) , "Handeln" (action) and "sinn" (sense). If we describe what people or animals do, without inquiring about their subjective reasoning for doing it, we are talking about their *Behavior.* If we study the subjective aspects of what they do, the reasons and ideas underlying and guiding it, then we are concerned with its *Sense.* If we concern ourselves both with what people are, overtly and objectively, seem to do (or not to do) and their reasons for so doing (or not doing) which relate to the world of meaning and understanding, we then describe *Action* (Weber, 1947 quoted in Reynolds, 1980, p. xxiii), but the translation is our own). Reynolds completes this by describing what an act is:

> "An action acquires its social meaning through being identified as a performance which when completed constitutes by convention an Act. Acts are not to be identified with either the actions needed to perform them, nor with the movements involved in the action. There are many different ways in which the same act may be performed . All have the same meaning through their identity with respect to the act of which they are the performance' (Harre and Secord 1972, quoted in Reynolds, 1980, p. xxiv)" (Johnson and Tomlinson, Systems Research and Information Science, 1994, vol. 6, p. 105).

Nevertheless, it seems desirable to stress the necessity that the outcomes of human activities in some way can be linked to some levels of measures (even if is it not possible to link some measure to basic research). Therefore it can be worthwhile to refer to the following: "From systems theory we also know that purposeful systems are comprised of purposeful subsystems which also require measures of performance. Some of these subsystems are themselves organizations, but, ultimately, an important primary component of any organization is its people. Thus, the measurement of human capabilities and activities is fundamental to measurement for a management decision. Similarly, organizations as systems are part of a larger suprasystem called the environment. An organization's environment is the source of its constraints or uncontrolled influences of its behavior. Consequently, environmental measures, i.e., measures at the social level, are equally fundamental for a management decision. The result is that three levels of measurement – organizational, individual and societal – constitute the measurement for a management decision" (Mason and Swanson, 1981, p. 23). In this connection it can be remarked that it seems difficult but not impossible to find some measurement for scientific research – at least in basic research. Maybe it can seem possible to find some measure for applied science , but the danger here is always present that the test of results will be unsatisfied . The researcher , namely, can be too eager to show his/her productivity .

Apropos the concepts of *purposive and purposeful,* it seems desirable to refer to Peter Checkland. He says: "I use the word 'purposive' when the meaning is the neutral 'serving a purpose' and 'purposeful' when conscious human action is involved. Thus the escapement is a purposive system of a clock ; telling the time by reading the dial is a purposeful action of a human being. So also are all actions involved in the monitoring and recording of our mental activity, enabling plans to be argued over, communicated, opposed, carried out, subverted, and/or modified in a purposeful fashion by groups of human beings. This purposeful activity distinguishes our species from the stimuli-provoked, goal-seeking, merely purposive behavior of the other animals" (Checkland, 1991, p.119).

However, even if the concept of behavior can be linked to various rituals and habits in different cultures , the model for the survey in question is connected to the overlapping cultural norms formed within the Western democracies. This report is also limited only to an overview of the three levels accounted for in the citation above, namely, the individual, organizational, and societal level. Such levels as groups, communities and supranational levels are consequently not objects of studies.

14.3. Behavior Related to the Level of Human Individuals

Apropos of knowledge engineering the following citation from Johnson and Tomlinson can be referred to: "From this perspective the knowledge engineer carries out knowledge aquisition activities to identify the acts which together constitute the expert's task. These acts may be instantiated by various actions, and the knowledge engineer has to ascertain the conditions under which certain actions are to be chosen in preference to others.

Modelling acts and actions, as opposed to behavior, requires that the knowledge engineer pays close attention to the expert's explanations for his or her choices." The authors says further: "Methodologies which model only behavior fail to give sufficient consideration to the expert's explanations. It then falls to the knowledge engineers to generate explanations from their analysis of the domain. But, as the knowledge engineer is a novice in the domain, building a system which operates at an expert level becomes problematic, if not impossible. Essentially, the engineer may know what the expert is doing, but doesn't necessarily understand it. This distinction between knowing and understanding has proved to be very important for conducting qualitative data analysis in the Social Sciences. It is a useful distinction for a knowledge engineer to keep in mind" (System Research and Information Science, vol. 6, p. 205).

The British Professor of Information Science, Frank H. George, gives his view of a human individual's cognition related to the concept of behavior. His view is illustrated in the following figures (Figures 35-37 or in his book Figures 1-3) There he has pictured the core of his thoughts of behavior connected to a behavioral and mental map of a person.

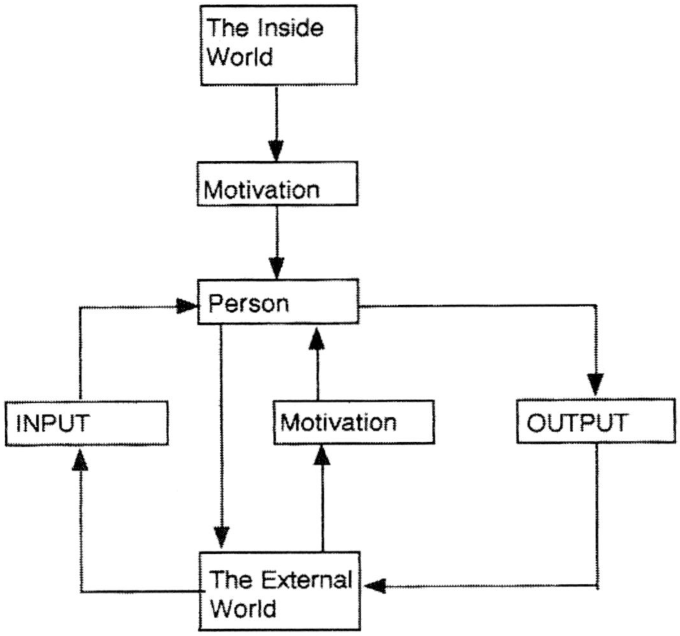

Figure 35. (Fig.1) A "behavioral" map of a person (from Frank H. George, Systems Research and Information Science).

He says among other things: "If we stuck with a narrow behaviorism of Watson or some of the other early Behaviorist we would find it difficult to discuss 'thinking', and Watson thought of it as 'implicit speech', similarly 'perception' was referred to as discriminatory reactions (Hilgard, 1948). So we accept that humans have both minds and bodies, but do not allow that the 'mind' is some etheral being within us. It is simply the working of the brain which deals with internal stimulus-response activity as well as the external stimulus-response.

If you like, Figure 35 (1) shows the strictly behavioral and Figure 36 (2) the so-called 'mental.' The combination of inside and outside worlds are featured in figure (3) 37.

The activities of the external world in relation to the person is clearly publicly observable, whereas the activities of the internal world are, in general not. All we can do is observe behavior subsequent to input and make inference about the structure and function of the inside world. The owner of the inside world can make statements about his awareness, his feelings, etc. and even describe the step-by-step process of a cognitive activity. That there will, from a scientist's point of view, be more uncertainty about this description of a private experience is obvious, since it fails to satisfy the scientist's usual standards of public observability. Against the above point it can be argued that scientists have for a long time now been prepared to make inferences about the unobserved (unobservable) and used theoretical terms to build a map of the internal (unobservable) state of affairs. Thus behaviorism in its later forms (Tolman; George and Handlon) has accepted an internal state and function which can be internally self-changing. So we can posit to theoretical terms such as 'belief', 'doubt', 'expectancy' and the like, and invoke these terms to make statements of a behavioral kind. So that we can deal with cognition without difficulty" (George, 1993, pp. 161-163).

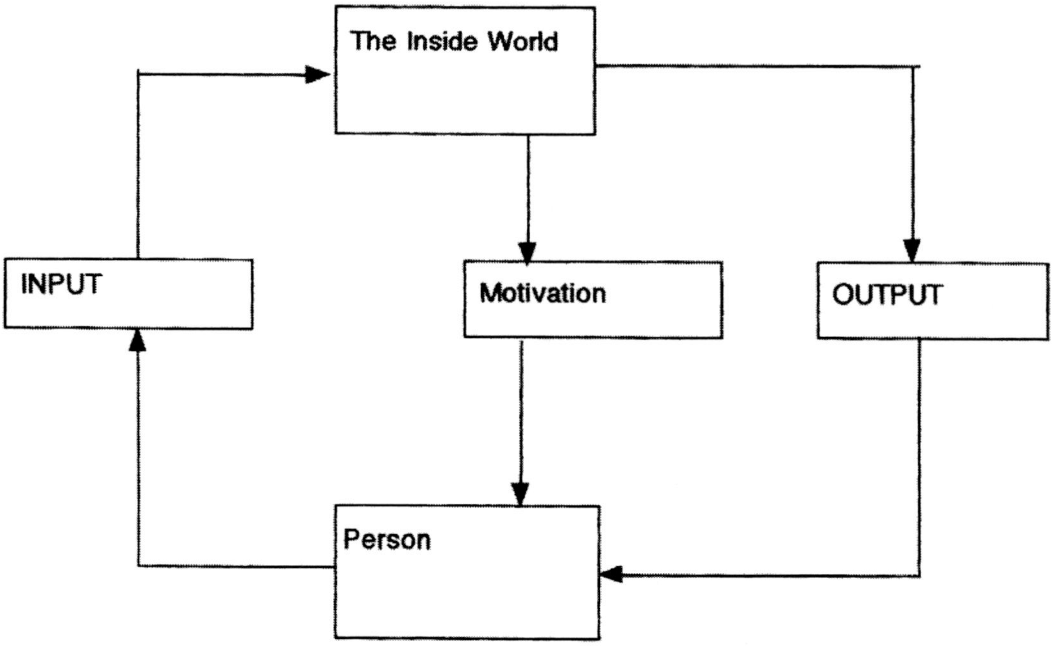

Figure 36. (fig.2) A "mental" map of a person (from Frank. H George, Systems Research and Information Science).

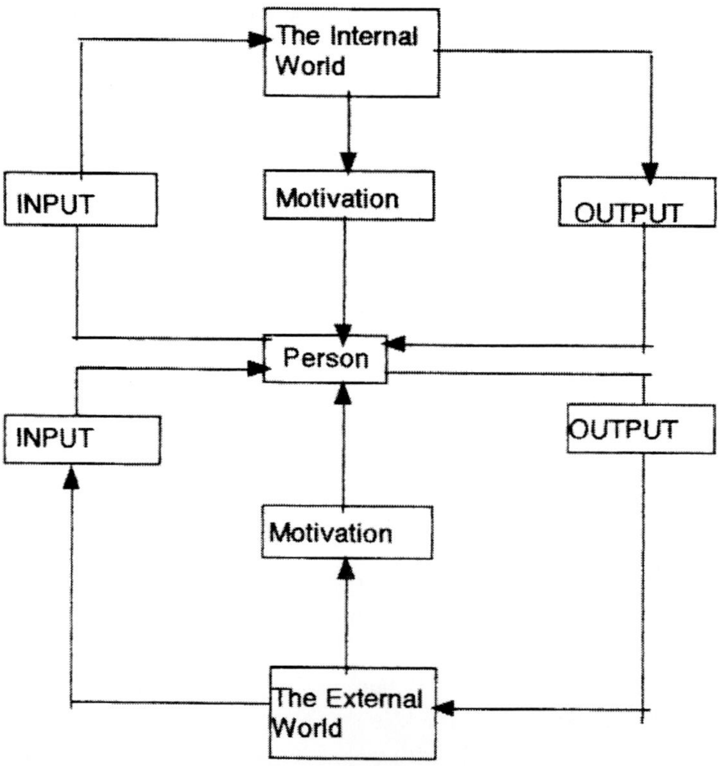

Figure 37. (fig.3) The behavioral and mental worlds combined (from Frank H. George, Systems Research and Information Science).

In this context it can also be of interest to refer to Max Delbrüch. He says: "In speaking of 'perception' and 'cognition' we are not referring to concepts extraneous to the physical world, implying the mind of a mysterious 'inner man'. Rather we realize that the beginnings of perception are already manifest in the microorganism, as represented by such phenomena as phototaxis in bacteria and phototropism in fungi. Even at this primitive level, the organism receives signals from the outside world, 'evaluates' their significance and responds appropriately. What do I mean when I say a bacterium or a fungus 'evaluates' the significance of signals? I mean no more, and no less, than the organism possesses structures that that permit him to react appropriately, that is, in an evolutionary adaptive manner. What do I imply by referring to a bacterium or a fungus as 'him' ? Do I refer to a subject with consciousness, with a mind? No, I simply refer to the organism as an indivisible, a functional unit and, as such, an individual"(Delbrüch, 1986, p. 273).

From the definitions given above it seems possible to state that the behavior of humans and animals foremost is dependent on inherited patterns of qualities, breeding, acquired qualities and influences from the environment. However, even if the human behavior is not as, George says, "some ethereal thing within us" but a function of the brain as a response to internal and external stimulus activities, it cannot go without saying that there is something that differs between the human minds and their products of behavior from the animals. This difference is the special faculty connected to the humans which lifts their behavior up above the animals, and that is the developed language together with the adaptive learning which can change operational plans for execution and break earlier conventions. Here it seems desirable to quote Klaus Mainzer. He says: "But only the human instinct of social organization for the purpose of survival is common with animals. Humans are distinguished by their political nature to achieve a just society. In a famous formulation, Aristotle said humans are defined as rational beings striving for truth in science and philosophy, and they are political beings striving for justice in society" (Mainzer, 1994, p.240). In this connection it seems also of interest to refer to the following: "Human social behavior is undoubtedly a legitimate candidate for biological study even if, when we study it as biologists, it eventually turns out that it is in fact so far removed from the principles that govern the behavior of other animals that present-day biological expertise can make no useful contribution. If biologists fail to understand human social behavior, it will not be because they are obviously doomed to failure as a matter of definition: 'because humans just are animals, and there is all there is to it'. No, if we fail it will be because we have no grips with the unusual biological complications presented by the pecularities of this particular species of animal, for instance language" (Richard Dawkins in *Science and Beyond*, p. 64).

But what is the basic prerequisite for expressions of behavior which lift us above other animals? The answer to this question can be sought in the following sentences; " What is certain is that the pre-adaptive conditions that allow us to assimilate language and culture is the morphological structure of our brains. Based upon studies of the relationships between increased brain size in homonids and the emergence of language and culture, anatomy professor Philip Tobias arrived at the following formula: Increase in brain size = gain in increase of neural organization = rise in complexity of nervous function = even more diversified and complicated behavior responses = progressively amplified and enhanced cultural manifestations" (Kafatos and Nadeu, 1990, p 96).

However, as Paul Davies says in his book, *The Cosmic Blueprint*, "It is important not to fall into the trap of supposing that all purposeful behavior is consciously considered. A spider weaving a web, or a collection of ants building a nest surely have no conscious awareness of what they are doing (at least, they have no conception of the overall strategy) and yet they still accomplish the task. The whole domain of instinctive abilities falls into this category . According to standard theory, the remarkable instinctive abilities of insects and birds are entirely due to genetic programming. In other words, nobody teaches a spider how to weave a web; it inherits the skill through its DNA" (Davies, 1987, p. 187).

At last it can be of interest to look at a citation from Karl Popper: "On all three levels of adaption (the genetic, the behavioral level, the level of scientific theory formation) adaptive changes always start from some given structures: on the genetic level, the structure is the genomene (the DNA-structure). On the level of animal and human behavior, the structure consists of the genetically inherited repertoire of possible forms of behavior handed on by tradition, and of open problems" (Popper and Eccles, 1981. p.133).

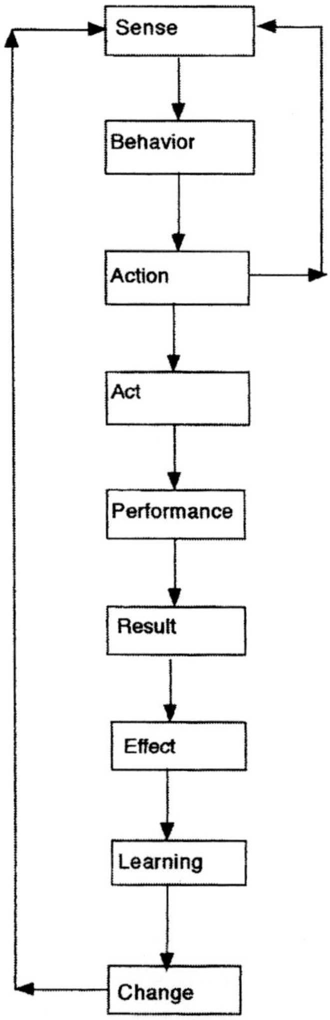

Figure 38. Components of the concept of behavior (from Rasegård, 1996).

Now it can be worthwhile to have a look at the following figure (Fig. 38). This figure is aimed at showing how behavior can be connected to certain expressions which can be regarded as fundamental contents of behavior mentioned in the foregoing and related to a human individual with the sense as governing concept.

Further we will try to link Professor George's Figure 3 (37) to the components of the author's Figure 38 letting them form a new figure (Figure 39).

Now we will connect this new figure to an example from the practical life – a person will be involved in the purchase of a house according to the following. A person (who we can call A) plans to buy a house. The thought of buying is originated as a product of his sense. That thought leads to activation of behavior. He decides to contact a house agent (who we can call B). He then has taken the first step of a chain of event linked to action. After consulting B, A decides to ponder the suitability of buying the house. This thought is passed back to his sense and the action is inhibited so far. After a time of pondering, a new thought of buying is originated as a product of the sense of person A. He decides to contact B again and then he has taken the first step of behavior in the second round. The house agent B is contacted and A has done his action for a second time. After this renewed action, A decides to act and buy the house. This buying means that he has done a performance. This performance gives the result that he is the owner of the house. But the result can have an effect which can mean that he discovers that the house is not of such good quality . This in turn means that he is learning something, and is likely to be more careful in the future, which influences his sense and changes his future behavior.

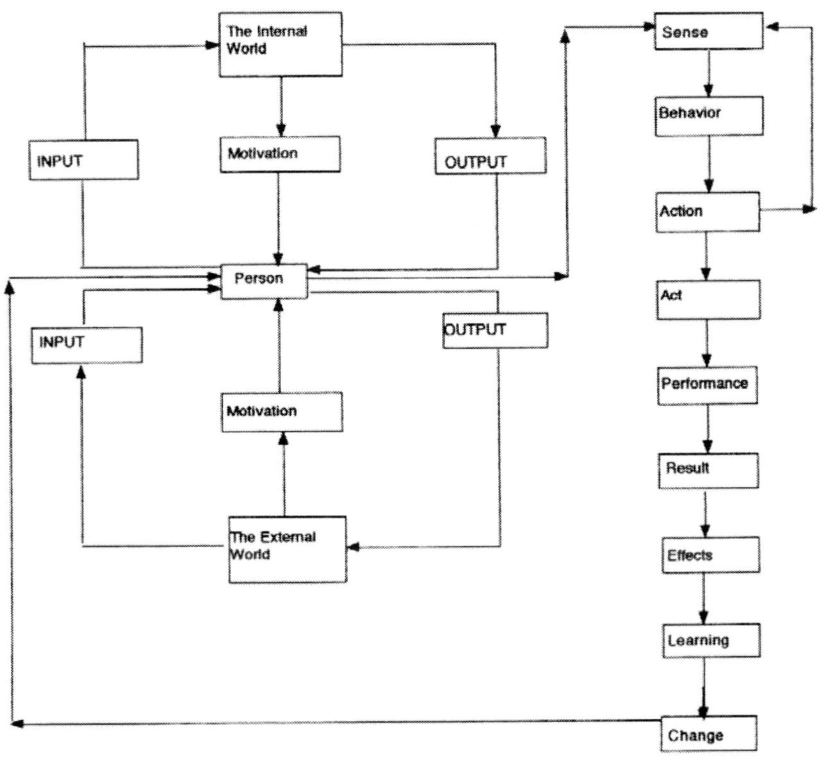

Figure 39. Linking man to a societal function (from Rasegård, 1996).

As we have seen, an action doesn't need to be followed by other expressions or measures. The action can mean that the act will not be fulfilled . However, if we decide to act we are deciding to do something with a special purpose, for example, the referred house buying. The act leads towards a performance, to produce something. The consequence of performance is accounted for as a result. This result has effects, both good and not so good. A good result of, for example, a company's performance can lead to the effects of improving the company's goodwill and that its financial status will be enforced. These effects have the consequence that the manager has learned something about how to sustain a good development of the company. This influences his sense and changes his further behavior. The circle is closed. An example of change at a societal level is given by Peter Checkland according to the following: "Probably the most spectacular recent example of change at a societal level which can be expressed in the concept of shifting appreciative settings, is the change in attitudes towards the environment and its conservation. The 'facts' about industrial depredation were available for many decades, but the readiness to notice them and evaluate them in a particular way seems to have been triggered by the publication in the early 1960s of Rachel Carson's book about the over-use of pesticides, *Silent Spring.* Thereafter the standard of what was acceptable changed sharply as new considerations became regarded as facts and new norms emerged. The rise of the conservation movement followed, and no industry or government can now ignore in its activity or legislation the care and maintenance of our small planet; the settings of the appreciative system has changed" (Checkland, 1991, p.263).

Concerning result and effect it can certainly be objected that both of these concepts have the same meaning. But here it is important to distinguish between the cause and the consequences. The result is simply an account for the outcome of certain performances while the effect is an account for how the result has influenced the learning in relation to sense and coming behavior . Therefore it seems necessary to draw a demarcation line between result and effect and see the effect in the light of a concept which gives the fundament of learning. At last it can be reasonable to devote a thought to Karl Popper's ideas of objective contra subjective knowledge. How can these ideas be applied to the example given above? The answer seems to be that here it is a question of some form of subjective knowledge, in other words, a knowledge acquired in our daily life of routines, consequently, not a knowledge on a higher level comparable with objective knowledge.

14.4. Some Roles Linked to Expressions of Behavior in an Organization

"In the initial stages of a study, the examination of elements of 'structure' and 'process' may be enriched by the examination of social systems as sets of roles with associated expectations of 'role holders,' behavioral norms and values by which that behavior is judged" (Checkland, 1991, p.18). Thus, in this section certain roles and values linked to expressions of behavior will be shown on the level of organization according to the following figure (Figure 40).

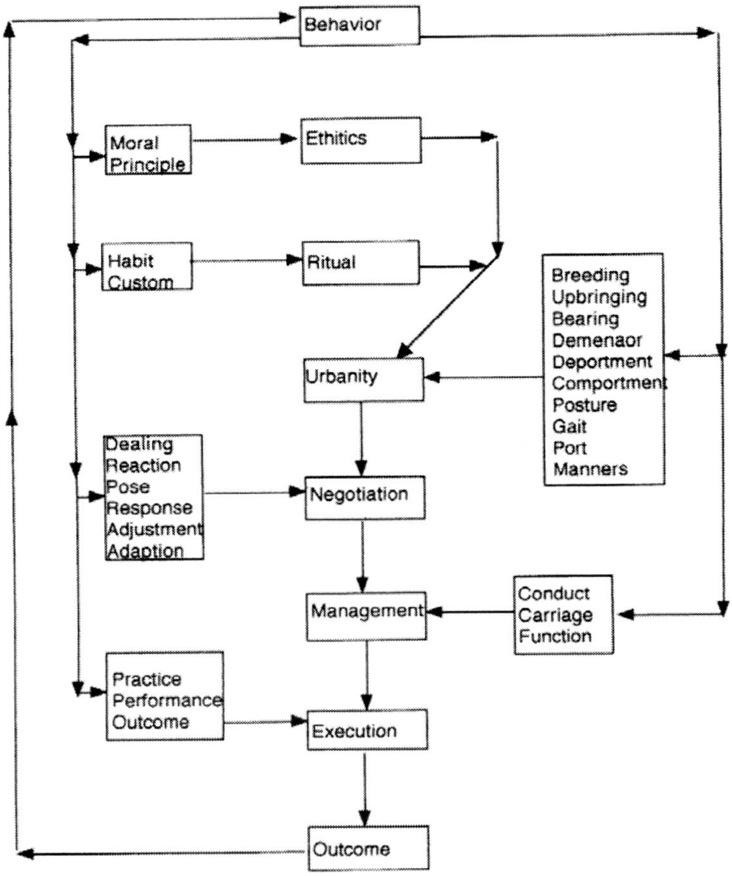

Figure 40. Expressions of behavior on the level of an organization (from Rasegård, 1996).

Certainly it can be argued that an organization consists of individuals and groups forming organizational units – sections and departments – and therefore this accounting for the relations of various expressions of behavior ought to have been done on a lower level. Nevertheless, the organizational level is the most complete object below the society level when accounting for the different expressions of behavior related to various plays of roles in the activities of a human association, for example, in the managerial role of a firm. There the manager, for instance, has to consider a couple of elementary behavioral rules of economic character in playing his role such as:

> "Rule 1 . A firm should not produce at all if the average revenue from its products does not equal or exceed its variable costs.
>
> Rule 2. Assuming that it pays the firm to produce at all, it will be profitable for the firm to expand output whenever marginal revenue is greater than marginal costs; expansion must thus continue until marginal revenue equals marginal cost"
>
> (Lipsey and Steiner, 1978. pp. 212-213).

And to play his role as manager or head of an organization he will be involved in other roles besides his principal task to supervise the daily activities in order to perform his duties according to the elementary economic rules given above, for example, negotiating with trade unions and suppliers. All of these roles are connected to the different expressions of behavior shown in Figure 42 and influence the organizational learning. Here it seems reasonable to give the following quotation from Kochen: "Learning behavior by a system consists in its giving to the same class of inputs or stimuli a new class of outputs or response, different from the class of responses given before. Learning responses may be primarily behavioral, consisting in directly observable changes in the memories stored within the system and applied to the internal steering of its externally observable behavior; or cognitive and behavioral learning may be combined in various ways. Thus a system may acquire new patterns of information that enables it to perform reliable operations of discrimination or recognition which it may not have performed before" (Kochen, 1975, p. 62). However, an organization consists of different human individuals whose actions are connected to fundamental components of the concept of behavior accounted for in Figure 40. And the roles shown in this section, of course all elements in the above-mentioned figure.

As can be seen in the figure, behavioral expressions connected to ethics play a superior role in relation to the concepts (or ought to at least). That means that a society without any form of ethical rules is a society in anarchy where the laws of the jungle are ruling. This ethical concept is foremost connected to a certain standard of moral and of fundamental principles of behaving in different situations. The sources of this ethical concept can be sought in religion and also in philosophical works of a various nature which have saturated the ethical thinking.

The rules of ethics are influencing the ritual behaviors manifested in habit and custom. The habit can be seen as an expression of the individual character with its roots in ethical thinking. The habit leads to a connection to custom which is manifested in the fundamental of acting within an organization as well as in a society formed by the rules within each organization as well as by the overlapping culture pattern ruling within the society.

From both foregoing groups of behavior, ethics and ritual, the group of urbanity can be traced. As can be seen, this group includes several expressions of behavior. All these expressions are closely related to a polished way of being and are forming the relations of human living founded on the ethical and ritual groups coupled to adequate breeding conveyed by parents and the educational system as well as voluntary organizations of education. Apropos the educational system, this system has an important duty that already in the basic schools convey a sense of morality and responsibility to the pupils.

The expressions within the negotiation group can be seen as actions which are necessary in our daily lives. In dealing with problem of negotiations we sometimes can be met by a false attitude from our counterpart, in other words, he can choose to react in an artificial way. But after we have adjusted our purpose it can be met with positive response and be accepted.

And now we are arriving at the management group. Here we can look at the conduct whose expression is of central importance when it comes to leading an organization as well as a society. The degree of conduct is important for the head of a firm as well as of a nation because it is essential to the carriage or, in other words, how well it is possible to carry on with the daily managerial affairs. The conduct and carrying on are of central importance to the function of the managerial unit, in an organization as well as in a nation, to the cooperation within the unit in question, as well as to its environmental relations.

And so we come to the latest group in this attempt of grouping the different expressions of behavior, the execution. Here is the question of how to practice the managerial intentions in the operation within an activity in order to obtain the outcome of a human entity. In this connection it can be remarked that both the expressions *action* and *act* are included in execution. The outcome of execution influences the chain of components shown in figure 40 with the individual sense as a superior element and can lead to a changing of the organizational behavior. In this context it seems possible to speak of a collective sense consisting of the individual sense of the organization's leader and his superior board.

Thus it seems possible to state that the different expressions of individual behavior are related to different manifestations of human activity in the society. But it is necessary to point out that the expressions of behavior are intertwined in each other. And that too much accentuation on an expression can contribute to the destruction of a desirable pattern of action, for example, a false position can cause a negative reaction of the counterpart in a negotiation and too much of adaption of a proposal can water-down the value of the original plan to next to nothing. As a conclusion of this section it can be suitable to point out the necessity of a behavior of cooperation between the individuals within the organization in order to reinforce the synergetic effect of work on all levels within the organization for gaining the most possible effectiveness, and here it can stand to reason to look at the following: "Biologist and Nobel prizewinner Sir Peter Medawar has drawn an interesting parallel between the emerging conceptual levels in physics and biology and the levels of structure elaborated in mathematics. In the construction of the concepts of geometry, for example, the most primitive starting point is that of a topological space. This is a collection of points endowed with only very basic properties such as connectedness and dimensionality. Upon this meagre foundation one may first construct projective properties, enabling the concept of straight lines to be developed. Then one can build up so-called affined properties, which endow the space with a primitive form of directionality, and finally a metric may be imposed that gives full meaning to the concepts of distance and angle. The whole apparatus of geometrical theorems may then be constructed. It would be absurd, Medawar points out, to talk of 'reducing' metrical geometry to topology. Metrical geometry represents a higher level enrichment of topology, which both contains the topological properties of spaces and elaborates upon them. He sees this relationship between mathematical levels in a hierarchy of enrichment as paralleled in biology. Starting with w atoms, building up through molecules, cells and organisms to conscious individuals and society, each contains and enriches the one below, but can never be reduced to it" (Davies, 1987, p. 146). So, here it seems possible to conclude that even the social entity which we denote as an organization with the described different expressions and groupings of behavior can not be reduced to individuals and groups but that the organization is acting as a body with different limbs which are supposed to be working together in harmony. In another case the organizational body shows a pathological feature and needs some cure in the form of restructuring. The author of this report has earlier in the American journal "Behavioral Science" given an example of an organization of a municipal nature where certain limbs of the organizational body were working without necessary coordination which caused the municipality in question (the city of Stockholm, capital of Sweden) a considerable economic loss (Rasegård, 1991).

Chapter 15

BEHAVIOR INFLUENCING THE SOCIETAL LIFE

15.1. EXAMPLES OF INFLUENCING SOURCES OF MALFUNCTIONS IN A CORPORATE "DEMOCRACY"

The risks of conformation of a democracy, as pointed out in previous sections, imply that one can speak of malfunctions of behavior on all levels of the human society. Some elements of behavior have been pictured in earlier figures. Nevertheless, it seems desirable to give a further completion in the next figure (Figure 41). Even if certain elements in the figure have been shown earlier, they here are inserted in another context in order to as much as possible convey a picture (even if it is sketchy) of the danger of conformity in a society with an originally democratic structure.

As can be seen, the figure shows two levels of decision, *determinator* or real decider and *decider* or formal decider. The real decider consists of corporate organizations who are representing special interests. The representatives of these organizations are not elected by the society's political voters and consequently are not a part of the democratic political power-structure. Nevertheless, they can be considered to have the real power of determining how the political decider function shall draw up its decisions. As can be seen, this determinator-function can be influenced by following four sources of influence.

Rasputinity influencer alludes to the Russian monk, Rasputin, who via the tsarina had a strong influence on the last tsar's political decisions. This influencer can be a mistress or lover of a powerful politician. But it can not always be a type of Rasputin or a madame de Pompadour. It can also be friends and relatives. Thus, it is a question of influence behind the official scenes.

Now we come to the *Corruption* influencer. In this case it is not a question of openly observed acts of bribery, no that is too clumsy. Instead, it is here a question of lobbying where certain persons want to promote their interests – political, economic or others – by influencing powerful politicians or persons who have close contacts with these politicians. It can also be a question of winning support in order to be appointed to an attractive position passing by other more qualified competitors.

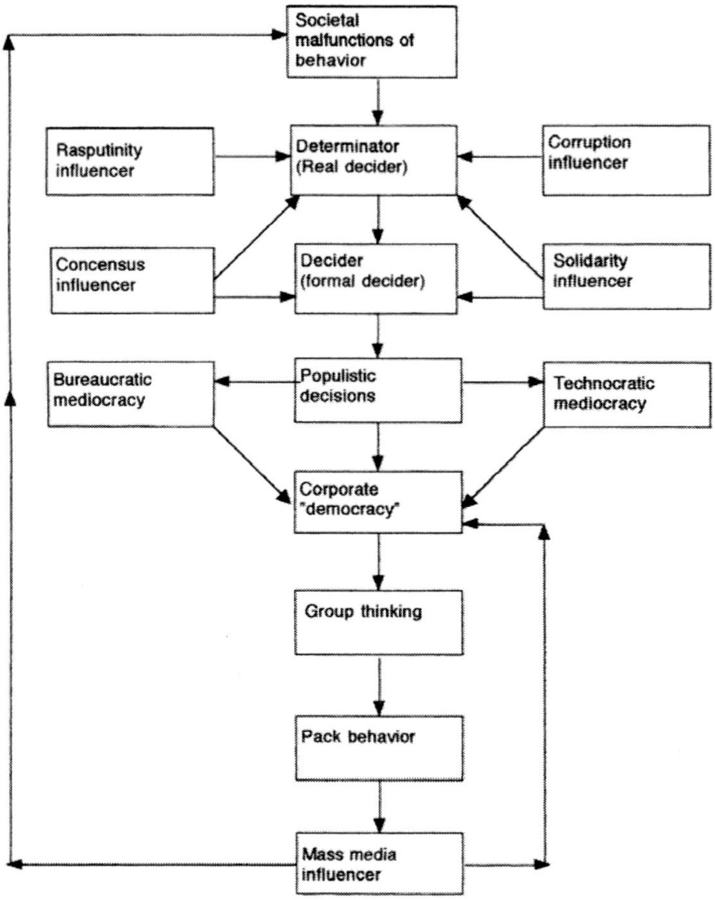

Figure 41. Societal malfunctions of behavior (from Rasegård, 1996).

The other two influencing sources, *Concensus* and *Solidarity,* have been accounted for earlier and need not be mentioned again. In the context of it can also be of interest to refer to some articles in a leading Swedish newspaper (Svenska Dagbladet) that uncover abnormalities in a Swedish organ of allowing grants for scientific research. The newspaper discloses that the members of this organ , scientists in leading positions, give themselves the double sums of grants compared to other researchers. Thus, the system has the consequences that the same persons who apply for grants are the same who allow the money for research. Because Sweden is a small country, it also means that for leading researchers and powerful politicians know each other rather well and the risks for friendship-corruption is obvious. For example the newspaper in question also mentions that only every fourth of new applicants for research grants received grants during 1994. Another example given by the newspaper is that a member of a research council for allowing grants, a Professor at the Swedish University of Lund, has received several millions of Swedish crowns for research without working on any research project at all. Not only did he receive these grants, he also stole other scientists' research results. In order to examine the affair, the research council appointed one of its members as an investigator. However, this investigator was in obligation to the council for received grants, had friendly relations with the accused professor and had been implicated in the council's decisions of allowing grants to him. These facts, of course, influenced his investigation. Later on it was discovered that another leading scientist of the council in question who applied for grants had an application with forged scientific enclosures. (Svenska Dagbladet 28th-29th June and 4th and 14th October 1995).

And so we come to the formal decider who has the political responsibility for decider actions. Here we will not see the Rasputinity or Corruption influencers because, as been said earlier, these influencers are working with the curtain down. The products of the formal decider's actions lead to populistic decisions aimed to satisfy the vociferous claims of rewards from certain political groups and organizations representing these groups' interests who in different ways can have supported the leading politicians to help them get their positions. However, these decision makers risk getting caught in their own trap because it can contain too much appointing of loyal party members in influencing organizations. And the competitors' loyality to the ruling party and near related organizations have been judged more valuable than applicants who have higher merits of education and experience. Thus, the risk is that it will result in the forming of a mediocre bureaucratic and technocratic cadre in the leading positions of officials . The mentioned factors lead toward something which can be called a corporate "democracy," where the concept of real democracy can be put within quotation marks.

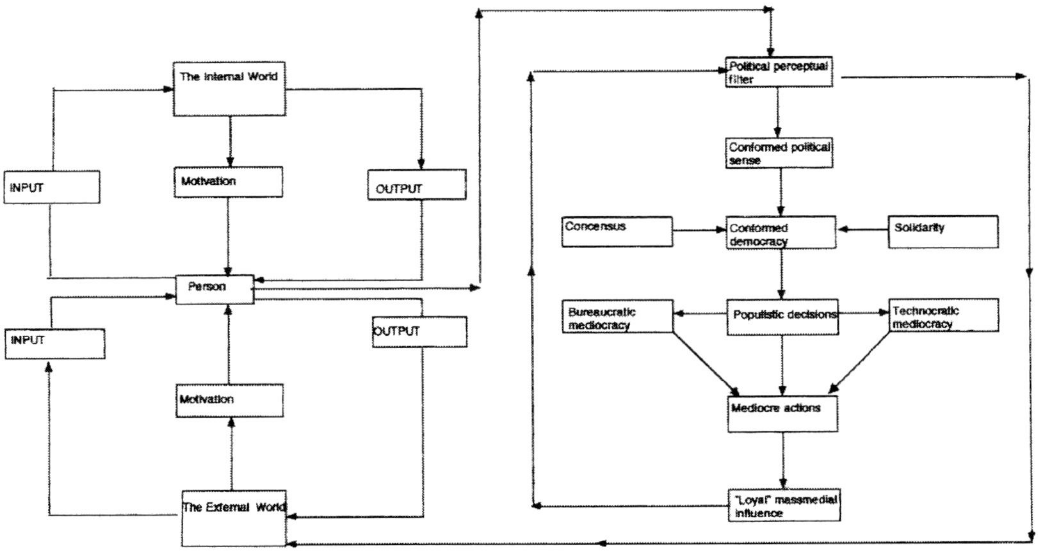

Figure 42. Societal conforming of man . This figure is a coupling together of Fig. 28 and Fig. 37 (from Rasegård, 1996).

The connected group thinking and pack behavior has been dealt with in an earlier context and therefore it seems not necessary to comment on these phenomena once again. At last we can see that the mass-media influencer is influencing and reinforcing the corporate "democracy." This is done by "loyal" journalists in press, television and radio but also by actors in certain films suited to give a favorable picture of the leading circles actions in the society and discredit the opposition. This influence also reinforces the societal malfunction of behavior in the way that the leading circles' decisions, even if these are wrong or not so good, are defended by these "boss-journalists." The above-mentioned elements of malfunction of behavior can under unlucky circumstances lead to a conforming of man featured in Figure 42 (Figure 42 is a coupling of Figs. 28 and 38.) Then we will face a brain-washing procedure equal to that practiced in a hard dictatorship. As can be seen in the following figure such a brain-washing contributes to reinforcement of the political perceptual filter.

15.2. Positive and Negative Behavior in the Societal Life

As can be seen in the sketchy drawing below (Fig. 43), we can distinguish two main branches of the concept of behavior . Both these branches have been dealt with earlier in this report. However, the four sub-branches featured below have not been mentioned .Therefore a brief description of these sub-branches will be given. Normal human behavior has been the main object of this report and is a behavior that common people use in their daily affairs. Idealistic behavior on the contrary is a more rare phenomenon . It is a type of behavior which does not have the goal of winning favors of some kind. Instead its goal is to serve mankind in a positive way without thoughts of rewards other than the satisfying sense of doing something good for humanity. As examples in our time can be mentioned the great French-German philanthropist and Nobel Prize winner Albert Schweitzer and his work as a physician and missionary in tropical Africa and also the British nurse Florence Nightingale and her work for an improved medical service during the last part of the nineteenth century. Certainly it can be argued that we in this context also have to include the saints of different times, but here it is important to remember that a great deal of these saints were beset by a religious fanaticism in their work and their aim was to collect merits for the heavenly world; their views of other religions, in many cases, can be presumed to have not been of a tolerable kind. The same can be said about the political fanaticism which sometimes can be a burden for humanity through intolerance toward other opinions. And now over to the branches of negative behavior. Here the defensive malfunctions of behavior can have its roots in some peoples' need to defend a political paradigm like what has been described earlier in this report and illustrated in an earlier context .This behavior has not as the main aim to damage but obviously it is a very destructive element in our Western democracy because it works for a conformity which undermines the ground for critical thinking and can, under unlucky circumstances, lead to a dictatorship. Now we come to the offensive malfunctions of behavior. As awful archetypes of individuals connected to this type of behavior can be mentioned Hitler and his extermination of the Jews, and during ancient times the Roman emperor Nero and his persecutions of the Christian people . Within a democratic society this kind of malfunction of behavior related to the political life has no place but is connected to brutal dictatorships. The behavior of exponents of the last mentioned type of behavior seems to be a behavior in which certain pathological features of malignity can be traced which maybe can be embedded in the genes of the individual from his birth – even if environmental influences such as good upbringing and education can prevent the evil features from breaking out.

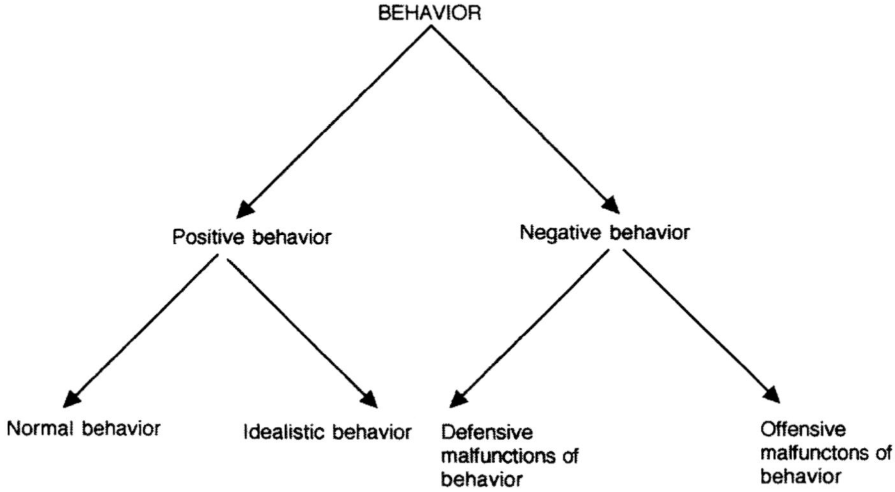

Figure 43. Branches of behavior (from Rasegård, 1996).

However, even if a country has a democratic political system, it cannot go without saying that within certain organizations and communities in this system certain abnormal features of behavior reminiscent of totalitarian systems' malignity can be traced. As an example can be mentioned the scientific community where it is not unusual that certain researchers can do their best to scandalize a competitor not as a defensive malfunction of behavior but as a pure evil destructive function in order to destroy a fellow-man's life and career. An example of this follows below. Another malfunction concerning researchers is that certain researchers tied up with political contacts can use these contacts to discredit a colleague in order to gain favor. In the Swedish journal Systemology, a journal of Informatics and Systems Science, the editor Tommy Kruszelnicki refers such a case. – A professor of social medicine at a leading Swedish university hospital, Huddinge Hospital, (Huddinge is a suburb to Stockholm, capital of Sweden) who had both international and national important commissions become more and more counteracted. The institution he headed which worked with interdisciplinary research was shut down and his research materials were destroyed. However, when this process was finished a new institution was established under another name, but with the same content as the former institution. This new institution, however, was provided with another head. The injured professor described this discrimination process as something of a Kafka process (Systemology, 2001, vol. 3.).

Another malfunction concerning researchers is that certain researchers tied up with political contacts can use these contacts to discredit a colleague in order to gain favor. Nevertheless, it can not go without saying that defensive and offensive malfunctions of behavior can be intertwined in each other to such a degree that it is difficult, not to say, impossible in certain cases to distinguish one from the other. The malfunction, for example, can begin with a defensive malfunction of behavior from a scientist in order to protect a ruling paradigm. This defensive malfunction of behavior then can be used by another scientist beset by an offensive malfunction of behavior in order to destroy as much as possible of the reputation of the exposed researcher.

15.3. Concluding Comments on Societal Behavior

The concept of behavior, as studied in this report, has been used as a background in describing the human actions in some societal functions. It is not the aim of the author of this report to serve as a lampooner of our Western democracy. Instead the aim is to point out certain tendencies which under unlucky circumstances can be enforced and lead to dangerous consequences of a governmental system which shall serve the people by free elections of their representatives in a free country - or in the words of the Russian-British actor Peter Ustinov: "Like most things in life, freedom has its good sides and its bad. Life is not worth living without it, but one of its bad sides is that life is not worth living either where freedom is not treated with the respect it deserves" (Ustinov, 1995). However, on the whole it seems possible to say that the Western type of democracy hitherto has served its purpose in a somewhat tolerable way compared to real dictatorships, even if certain serious imperfections and dangerous tendencies from time to time can be uncovered. Therefore it seems not necessary to go as far as Winston Churchill who has been ascribed to an utterance that "democracy is the best of all bad governmental systems."

The cardinal question here is the concept of behavior and its expressions in different societal functions where the ethics play a superior role in combination with critical thinking and observing. Apropos both of these verbs it may be reasonable to as a curiosity to mention following anecdote: "I can only finish by telling you an apocryphal story of Wolfgang Pauli, the famous theoretical physicist who suggested the existence of the particle we have never really detected – the neutrino – which has no mass, no charge, almost nothing. When he died he went to Heaven, as all good theoretical physicists do. When he was in the Presence he was asked: 'What would you like to know?' So he said, 'Well, how does it work?' 'Well, said the Presence, 'It would take a long time to explain it.' Pauli said, 'I've got all the time in the world,' so the Presence stated and he covered blackboard after blackboard – finally on the 7th day he collapsed in a cloud of chalk and said: 'What do you think of it all?' Pauli said, 'Umm, very interesting – but weren't you wrong by a factor of 2 on the second line?' 'You are right' was the astonished reply. Naturally the response to Pauli's critical approach was not unimportant" (John Taylor in Science and Beyond, 1986, p. 167).

Maybe it can be reasonable to finish these concluding remarks by repeating the following reflections by Paul Davies despite their having been cited in an earlier connection: "I have been at pains to argue that the organizational principles needed to supplement the laws of physics are likely to be forthcoming as a result of new approaches to research and new ways of looking at the complexity in nature. I believe that science is in principle able to explain the existence of complexity and organization at all levels including human consciousness , though only by embracing the 'higher level' laws. Such a belief might be regarded as denying a god, or a purpose in this wonderful creative universe we inhabit. I do not see it that way . The very fact that the universe is creative, and that the laws have permitted complex structures to emerge and develop to the point of consciousness - in other words, that the universe has organized its own self-awareness – is for me the powerful evidence that there is 'something going on' behind it all. The impressions of design is overwhelming. Science may explain all the processes whereby the universe evolves its own destiny, but that still leaves room for there to be a meaning behind existence" (Davies, 1987, p.203). In this last perspective it can seem futile to occupy one self with such matters as searching for malfunctions of behavior and malfunctions of democracy. Nevertheless, the expressions of behavior form our culture and make it possible to continue with our scientific and cultural development as well as a well-functioning democracy is necessary in forming the human individuals into individuals of critical thinking which is the fundament of all civilized development.

Chapter 16

PRESENTATION OF A WELFARE SYSTEM

16.1. WHAT IS A WELFARE SYSTEM?

Because a welfare system is a profound element in a political democracy it seems desirable to give at least a glimpse of problems connected to this complicated system. This report's aim is to focus on central functions as well as malfunctions of a modern welfare system, mainly based on the Swedish model. It can be said that a primary goal of a welfare system is to protect people from destitution regarding the common standard of living of the country in question. That means that what is considered as a satisfying standard of welfare in a less developed country can be considered as utterly inadequate in a rich country. Thus, it seems possible to speak of welfare of various degrees. Even if it has been said that the primary goal of welfare is to protect people from destitution this can be seen as too limited a goal as seen from a wider perspective. Welfare, namely, must mean a good care of old people and include pensions that give possibilities for pensioners to keep a satisfying standard of living. The working people must also have satisfying salaries and wages so they can afford a good standard, for example, of nourishment, housing and possibilities of recreation. However, not to forget, it is very important that all people have access to good education and good medical care. Another important ingredient in a welfare society is that the citizens have access to cultural activities and possibilities to use these activities in full. And that will mean that the earlier mentioned components of a welfare system must be satisfying. It is difficult to appreciate cultural activities when one is tired and the purse and stomach are empty.

16.2. MEASURING THE DEGREE OF WELFARE

Thus, on the whole one can say that a welfare system is a very large and complicated concept. Therefore it is difficult to obtain a common measure of the standard of welfare. Certainly it is possible to construct a common measure by weighing together indications from the different branches of welfare. However, there will always be different opinions of the importance each measure ought to have in a common scale. Another way to measure the degree of welfare between different countries can be to compare the costs of welfare in relation to each country's total expenses of the total budget. However, that can only be a very rough measure considering the cost effectiveness of the delivered welfare. In that question it also needs to be considered if the costs of the welfare can be considered as (normal) adequate in relation to the spent resources. And that will mean that the country can have a relatively high percent of welfare costs in its budget depending on a lesser level of effectiveness in the production of welfare services, which follows of wasting resources, both personnel and others. Thus, it must be a question of certain qualitative measures of each branch of the welfare system, for example, queue times for medical care in hospitals and other medical treatments. But even such a measure is not relevant without a qualitative measure of the treatment in question which can be rather complicated. Only to give a measure as the numbers of treated patients without coupling to the quality of treatments are not enough. In this context, it may be can be of interest to look at the following citation:

> The decision-making problems of any of the aspects of measurement are enormously difficult, and even an approximation to their solution still escapes us. Everything that has been said here about measurements is applicable to a broader class called "information" and "data". A rather significant portion of our resources is devoted to generating and processing data. However it is apparent that no one knows how the data should be expressed (the decision problem of data *language* is unsolved), what data are needed (the decision problem of data *specification* is unsolved), how the data are to be used in various contexts (the decision problem of *standardization* is unsolved) and how the data are to be evaluated (the decision problem of *accuracy* is unsolved) (Richard O Mason and E. Burton Swanson, 1981).

So, even if the use of reliable measurements have been improved since the citation was written considering (among others) an improved computer technique it can be of value to think about the problems in the citation above.

However, as a minimum of accounting for welfare costs it seems desirable that the authorities at least give a specification in the demand note of the cost of the different branches of welfare in order that the taxpayers get at least a rough idea of what they pay for. And that is often done in an enclosure to the budget. Finally can be mentioned that a condition for developing and improving welfare is a well functioning infrastructure, for example, roads, railways and other communication means. A good infrastructure, namely, is the backbone of welfare.

16.3. Two Main Levels of Welfare

The diagram below shows the different components of a welfare system. At first we can see the connections to the infrastructure. A good infrastructure, namely, is a prerequisite for developing a welfare system. As can be seen the welfare system demands improve of the infrastructure. On the other hand an improved infrastructure contributes to the development of the welfare system. Hence the arrows to and from the box above. Further we can see that it is possible to speak of two main levels, namely, elementary and supplementary. The elementary level includes the most important components for human survival without suffering individual destitution. To begin we have food and clothing representing necessary nourishment and a sufficient standard of clothing. Concerning housing it includes a good standard of buildings for common people at reasonable costs inclusive to a certain officially determined minimum area per individual. As can be seen from the diagram even education is included in the welfare system because educated people are a necessary component to keep the welfare system going. Without an adequate educational system, the base of the welfare system is in serious danger. In other words, it is not possible to develop and keep the welfare system intact. Another very important ingredient in a welfare society is medical care with good access to medical centers for preventive treatment as well as hospitals for acute and more serious illnesses. And so we come to other attributes. This concept can, for example, include telephone, radio, TV, simple communication means and good access to common transport means.

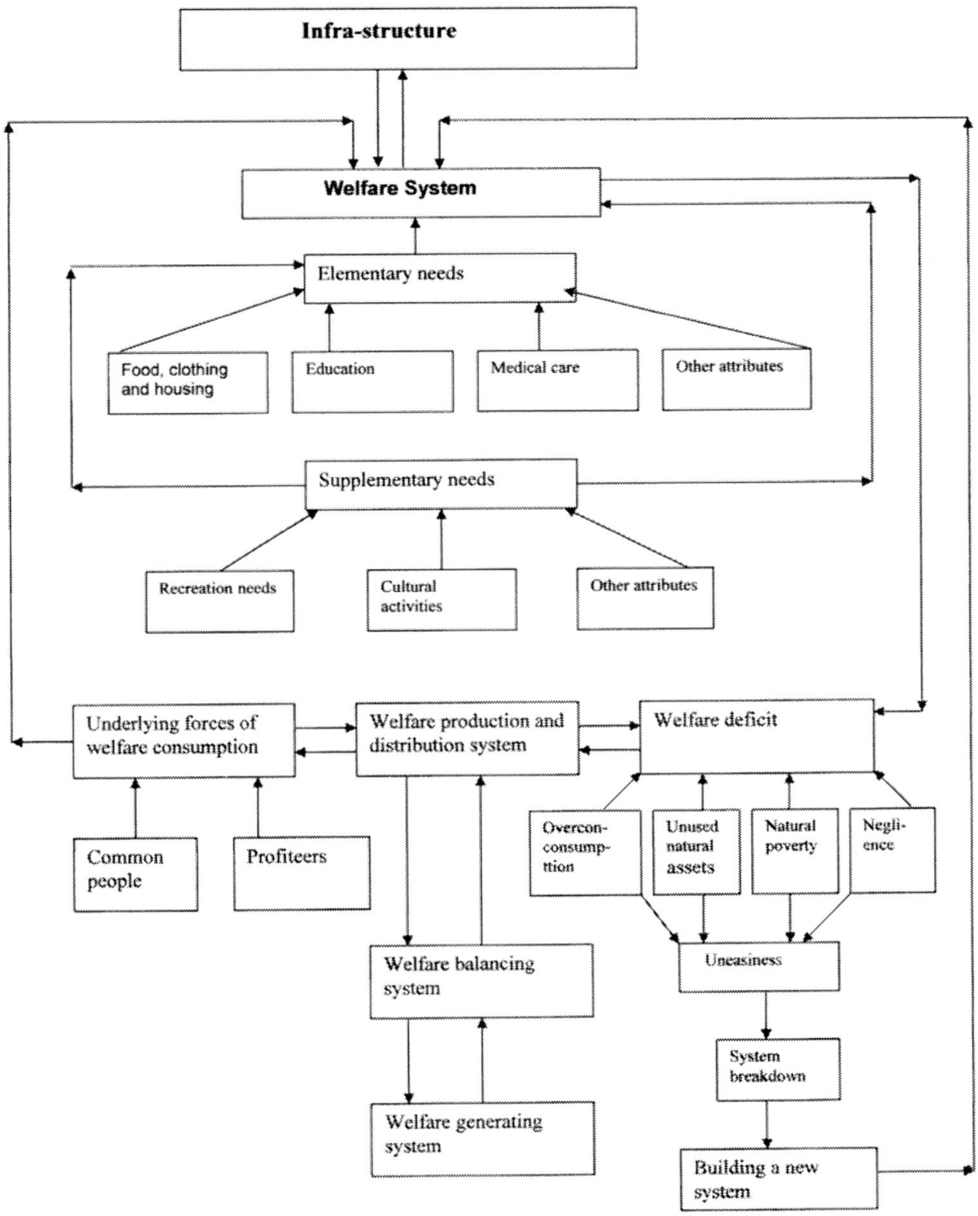

Figure 44. Basic components of a welfare system.

And now over to the supplementary level which comes into focus when the elementary needs are satisfied. Even if these components can have a position above the basic needs they are important for keeping and developing a healthy and culturally active people. Here access to good recreation activities are important for healthy inhabitants and to prevent diseases. The access to cultural activities is a necessary attribute to develop a civilized society characterized by humanity and tolerance. Considering the other attributes of a welfare system it can be a question of using, for example, motorcars, sailing boats and sport cottages etc. But all are relative. And after a country becomes richer with a higher standard of living these other attributes gradually can be included in the elementary needs. However, it is also to be observed that the demand for quality in all these components, both elementary and supplementary, will arise gradually as a country becomes richer and at the same time increase the quality of life. So, it is here a question of a dynamic process over time.

16.4. Demand Machine

Now we can have a look at the underlying forces of welfare consumption, the so-called demand machine. Here we have the collective demands from common people according to the components accounted for above. However, within a welfare system it is always profiteers, rent-seeking individuals who profit from the system in order to get favors which they don't have the right to claim, for example, faked sick absence, work at the black market while receiving unemployment support and using other methods to cheat the welfare authorities. These profiteers can be denoted as the welfare system´s grave-diggers. The two mentioned groups of consumers, the common people and the profiteers, are inspired foremost by the mass media and trade unions and not in the least by the authorities and officials within the welfare system. The last source gives special information on the right to claim which the profiteers use for their dishonest activities. What is said here will be further elucidated later on in this report.

16.5. The Base of Performances

The next component that has to be studied is the base of performances by the welfare system. As can be seen in the diagram, we have to consider three main components - the production and distribution system, the welfare balancing system, and the welfare generating system. First we can have a look at the production and distribution system. This system is serving all the main activities included in elementary and supplementary needs shown in the diagram above. The next system, the welfare balancing system, is a collective concept of all political processes on all levels of the society that have the responsibility for financing the welfare and deciding the rules for welfare distributions to the citizens. And so we come to the welfare generating system or, in other words, the main base for a functioning welfare system. This last mentioned system contains all the individuals and organizations that are providing services and products for use in the trade internally in the society and not least externally for exports. Even scientific research activities aimed to develop the base of welfare are included here.

As can be seen from the arrows, the underlying forces influence the welfare production and distribution system to produce and distribute more welfare which then will be consumed by these forces. At the same time we can see that the production and distribution system tries to eliminate or decrease the deficit, but that will also mean higher demands of the welfare production and distribution system. This in turn will mean a weakening of the whole welfare system. These reciprocal influences are indicated by the arrows to and from the production and distributing system.

Concerning the arrows to and from the welfare production and distribution system, they indicate that this system produces and distributes welfare according to the demand of the welfare balancing system.

At last we come to the welfare generating system. This system receives the demands in the form of charges and taxes from the welfare balancing system and delivers the necessary financial means which can be called the financial base of welfare. This is indicated by the arrows to and from the welfare generating system.

16.6. Welfare Deficit

Earlier certain factors that influence the consumption of welfare had been mentioned. In the boxes to the far right in the diagram certain causes of a welfare deficitare pictured. First we have over-consumption when a country is living above its financial resources. An example of this is given later on in this report in connection with an analysis of the base of performances of the welfare system. Another source of welfare deficit can be unused natural assets in which the country does not have enough financial means to exploit in a more large scale way. It can also be that a country can have something that can be denoted as natural poverty, or in other words, it has meagre natural assets and the people´s living is mainly based on a primitive agriculture. At last but not least we have to consider the negligence, governmental mishandling, which can not be neglected even in highly developed countries – a waste of resources, personnel and others, as a result of ineffectiveness. The causes of welfare deficit that are mentioned can be sources of uneasiness and lead to a system break-down. However, the chaos of the system cannot continue forever. Gradually a new system will take its first steps toward a system of welfare. The circle in a dynamic process is closed.

16.7. Welfare Deficit Forever?

On the whole it seems possible to say that no country is doomed to misery forever. Beginning with the over-consumption and the risk of misery of overspending, the mass media connected to a ruling party or a coalition of political parties has the responsibility of not giving a too rosy a view of the actual economic situation. The same concerns the trade unions. The author of this report has analyzed certain influences in this connection by political and other factors in his book "Man and Science – A Web of Systems and Social conventions," (Rasegård, 2002). And now over to countries which have good natural assets but not exploited these in full. Here it can be a question of a poor educational system combined with unwillingness to let investors from other richer countries exploit the natural resources – it can be pure patriotic reasons combined with self-esteem and frightened neo-colonization. In this case international funds, for example, the World Bank ought to have a mission to support the country in question with advice and money for developing the resources. Considering the natural poverty of a country with poor natural assets and a people who have a primitive living based on cattle and primitive agricultural activities, that means that the standard of living needs not be unchangeable. It is not impossible to lift the people to an acceptable level of living through education and training and in that way stimulate the citizens to improve their agricultural methods and create new ways to form a living base, for example, developing the tourist industry. It will here be a question of financial aid from international organizations, e.g., the aforementioned World Bank. However, in all financial assistance of underdeveloped countries, it is important that a main part of such assistance is aimed at self-help. In another case it can be like public assistance of a social welfare character given without demands on performance – or like feeding open-mouthed nestlings in a bird-nest.

At last we come to the governmental mishandling of resources. Here it can be a question of a corrupted governmental system. In this case it will be a duty for politically conscious people to act for reforms and improvement in the way that the country ´s potential for an effective and good welfare system will be used in full.

Chapter 17

A Welfare System Seen from an Analytical Viewpoint

As hinted at before, an elementary prerequisite for science to develop and prosper in harmony and give benefits to a society is that it has been given enough financial resources. That means that the society (the country respectively) can stimulate its citizens, the scientists and others, to be productive and at the same time enjoy the rewards of their productivity. Therefore it can be of interest to look at the following viewpoints of some dangerous features in a social welfare state. Of course, certain scientists, perhaps, will shrug their shoulders and say that lies out of this description of man and science. Nevertheless, scientists are also depending on a well functioning welfare system which also has enough resources for their research. And here it also can be a question of political decisions, if decision-makers are giving priority to some social welfare project and reducing some of the allowances for scientific research, for example, ecological research.

17.1. Main Analytical Components in Brief

As an illustration of the main analytical features of a social welfare system the following figure (Figure 45) can serve. (The comments have the Swedish situation as the model).

The political decider system is represented both on the state level (parliament and government) and local levels (deciding assemblies of counties and municipalities). On the state level, taxes for overlapping social welfare are decided, as well as subsidiaries for certain welfare activities on a local level. On the latter level, the local tax share of the welfare expenditures is decided, as well as charges and tariffs of certain services within the social welfare area. Even if these local charges and tariffs are connected to some types of social welfare, it is important to remember that even these types are more or less subsidized by the state or local communities or both.

The welfare financing system is formed by a conglomerate of financial sources of different types (on the state and the municipal level), taxes, charges and subsidies from the State to the municipalities for certain welfare policy, e.g., housing policy . An important role is played at present by loans from internal as well as external finance institutions.

The welfare administration system consists of authorities and officials employed for administration of the social welfare system. These authorities and officials are placed both on central and local levels.

The welfare producers system is a network of employees in the state as well in the local welfare institutions who are directly involved in service work within the social welfare production, e.g., social care workers, nursing staffs at day care centers and other personnel employed at welfare institutions.

The welfare consumers (demand machine) consist of all citizens who enjoy social welfare and who demand services of different kinds, revenues for taxes or, in other words, makes it possible for the welfare system to exist and to develop.

The welfare generating system consists of enterprises in the society's trade and industrial sections who produce goods and services whose income form the basis of revenues for taxes or, in other words, makes it possible for the welfare system to exist and to develop.

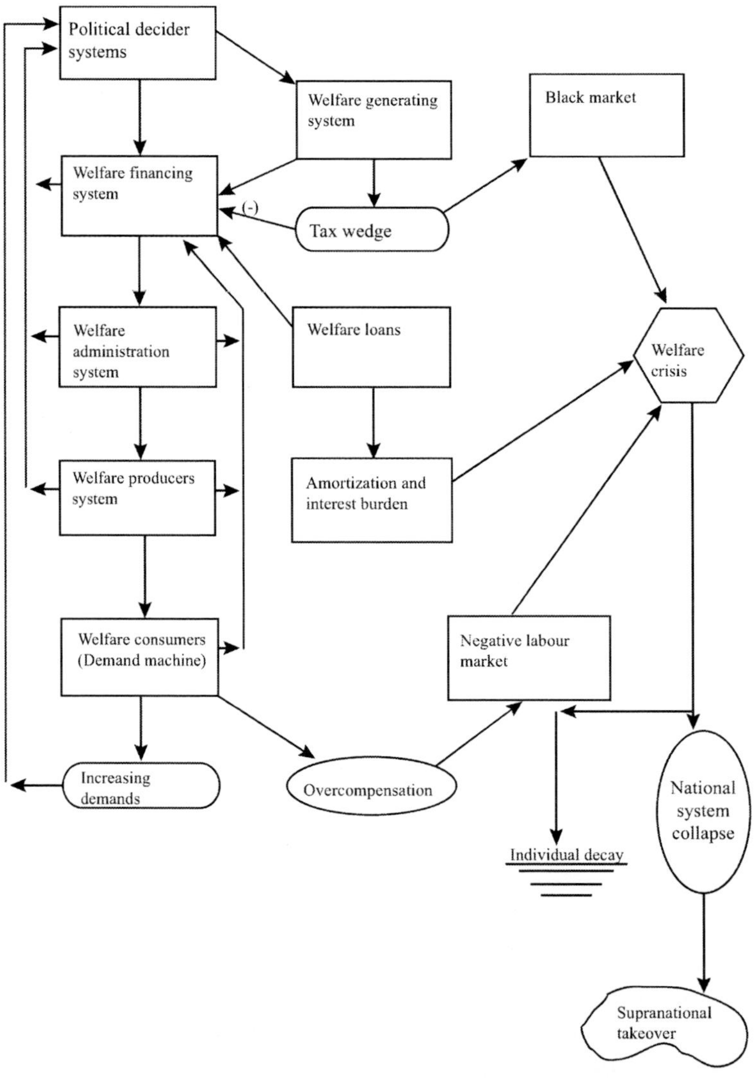

Figure 45. Main components of a welfare system from an analytical viewpoint (from Rasegård, 1996).

17.2. An Analytical Look at the System's Functions – or Malfunctions

And now for an analytical look at the social welfare system. As can be seen from Figure 45, the political decider system determines the tax imposed on the welfare generating system . Here it is a question of taxes of different kinds, even such taxes which are denoted as "charges," e.g., the employer's social charges to the state calculated on the basis of the employee's wages and salaries. However, it may seem remarkable that personnel employed in the system contribute to financing, which the arrows on the right of the system boxes in the figure indicate. Nevertheless, these employees have to pay taxes for their salaries and wages of which an amount is conveyed back to the financial system. In the same way it is possible to say that even the social welfare consumers contribute to financing because pensions and other income emanating from the social insurance is taxable income. Besides that, certain welfare consumers have to pay subsidized charges for welfare service. Thus the figure in question indicates a gross account of social welfare financing. Therefore, it is possible to state that, in this case, there is a certain going round of money in the system.

When we look at the individual tax burden, we ought to pay attention to the tax wedge which is the sum of taxes as a percent of taxable income. This measure indicates the percent to which an addition to an income generating effort will result in an increased tax. As can be seen, the tax wedge can be of such a size that it can diminish the inflow of money to the welfare financing system. This high tax wedge encourages the tax swindlers and an increasing black market which is avoiding taxation. Also, citizens who have the ability to use every loophole in the tax legislation can be encouraged to investigate every possibility of tax-evasion and the judicial tax experts will have a golden market. The tax morality will be eroded – it does not seem bad to cheat the tax-authorities.

The just mentioned factors lead to the fact that the welfare generating system cannot mobilize enough economic resources for financing the social welfare. The deficit must be financed by loans. And that means that the coming generation has to pay for the present generation's overconsumption in the form of amortization and interest burdens. Another fact which can be dangerous is that too high of an economic compensation of unemployment and other social benefits can weaken the incentive to work. In other words, the danger is that an army of individuals unwilling to join the regular labor market can be formed. Then we will be confronted with something which can be denoted as a negative labor market. These individuals may then be seen as parasites on their fellow workers in the regular labor market or, in other words, we can distinguish the rise of a new exploiting class of unemployed citizens which lead thoughts back to the old feudal system.

17.3. An Alarming Scenario

Going back to the welfare administration and the welfare producer systems, both systems can also, aside from engaging themselves in the matters of the social welfare policy, be considered to have a pure egoistic interest as employees within the system. Of that follows that they have an interest in defending the public sector of society because they earn their living within this sector. Therefore it seems possible to talk of a so-called "welfare-mafia" who defends and, if possible, wants to enlarge the public sector. Because of that, this underlying desire foremost is connected to the socialistic parties with connected trade unions. These parties and organizations have the dominating responsibility of not damaging the welfare generating system to such a degree that it loses its ability to compete in relation to foreign industrial organizations. And this may result in the fact that successful national enterprises might be tempted to move out of the country . This can be completed with an increased black market together with a negative attitude to work (a negative labor market).

Now we come to the terrible end of this scenario and that means that the national social welfare system will collapse and a supranational organization, e.g., the World Bank, will take-over and dictate the economic policy in the country, in other words, the country has lost its independence. Besides what has been said, the risk is that the national system collapse leads to individual decay and morality will be wiped out. That means that a former prosperous society will end up in poverty and anarchy with a high degree of crime - the law of the jungle will be ruling. Another danger is that the social authorities in their eagerness to control the claimants of welfare will tighten more and more the control of citizens by developing interactions between different computed data systems, and the society runs into a network of social control. In other words, the citizen in the democratic society will be exposed to something which has certain features of the control exerted in a totalitarian state.

However, in the welfare society it is always a danger that the citizens can be encouraged to be rent seeking individuals who all the time are hunting social benefits. What to do to avoid such a situation? An answer to this is that the social consciousness aiming to reinforce the individual's responsibility for his life should be conveyed by the educational system and not in the least by the forces who are connected to the trade unions. However, here the political parties have a big role to play when avoiding a costly overbid policy in social matters without thinking of society´s financial situation.. Nevertheless, the social welfare and the foundation of its existence is ultimately a question of the individual's responsibility and solidarity toward the welfare society.

17.4. Some Components in the Process of Forming a Welfare Opinion

As can be seen in the figure below (Figure 46), we have to consider three main paradigms of social welfare. These paradigms form together an overlapping paradigm generally accepted by the welfare authorities and their employees, and within this paradigm they have to do their work.

The features of these three main paradigms as constitutes the overlapping paradigm can be described as follows:

First we can have a look at *the welfare ethical paradigm.* This paradigm has in western democracies been influenced by Christianity's theme of charity. A main source of influence has also been the socialistic ideas brought into practice by the labor movements . But also social-conservative and liberal ideas of a society's duty to give assistance to exposed people have played an important role. A leading thought within this paradigm is that nobody shall be

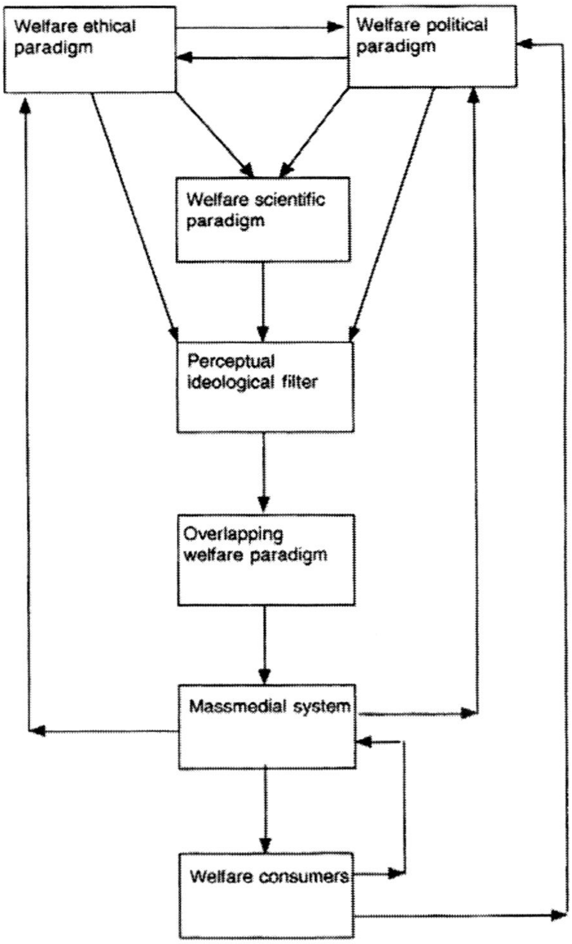

Figure 46. Components forming a welfare policy (from Rasegård, 1996).

exposed to destitution and that all citizens shall have the human right to a reasonable standard of living. Now we can continue with a study of some basic features of *the social political paradigm*. This paradigm is connected to the political decision makers' views of social welfare. Here the political party or constellation of political parties who are in office form the ruling policy of the content of social welfare. The Swedish journalist and social debater, Anders Isaksson, has in his book "Alltid mer, aldrig nog" (Always more, never enough) criticized certain features of the Swedish welfare state. He among others refers to the sixth paragraph in the Swedish Social Services Law which can be translated into English according to the following:

> "The citizen has the right to assistance from the Social Board for his/her support and way of life, if his/her needs can not be provided in another way. The citizen shall through the assistance be guaranteed a reasonable standard of living. The assistance shall be so designed that it reinforces his/her resources of living an independent life."

Isaksson's comments on this paragraph can in suitable parts be translated into English as follows: " In this legal term, three short sentences, which in 1980 was issued by a non-socialistic government, states the unconditional right to a reasonable standard of living – irrespective of whether he/she is working or not. The entire right, consequently, is unconditional and cannot, in consequence of that, be coupled to demands of any form of equivalent performance done by the receiver, for example, doing some social service work during some hours per day (Isaksson, 1994, pp. 108-109). Thus, the leading thought within this social political paradigm is the citizen's unconditional right to a reasonable standard of living free from any equal contribution to the society.

Now the *welfare scientific paradigm* comes into focus. This paradigm constitutes the social engineering within whose framework the researchers work out different models of further development of the welfare system as well as strategies for their implementation. This paradigm is protecting the ideas of social welfare from such ideas which can be judged as dangerous and therefore threatens the leading political view on the welfare policy, for example, increased private services in certain sensible parts of the social system as children and hospital care as well as social and medical care of elderly people. The researchers within this paradigm often have the task to array the decision makers' decisions in some sort of scientific disguise. Thus, the leading thought within this paradigm is that the society shall have the responsibility for the main social welfare and this responsibility shall not be transferred to charitable organizations or private interests.

From the three mentioned paradigms, the information passes through an ideological perceptual filter or, in other words, the information is more or less colored depending on the ruling ideas within the ethical, political, political and scientific welfare paradigms. After this filtering, the information is incorporated as a part of the overlapping social welfare paradigm, accepted by the welfare authorities and their employees, for further conveying to the mass-media system with this system's shaded meaning and political inclinations when forming public opinions. As known, the mass-media system is a forceful power when it comes to exerting influence both on the ethical as well as on the political welfare paradigm. Through the mass-media system, the welfare consumers are informed about actual questions of the welfare policy. This information forms the opinions from the welfare consumers which in turn are conveyed to the mass-media system through their organizations, e.g., pensioners', tenants' and parents' associations and others representing the receivers of social welfare, and last but not least the unemployed members of trade unions.

Now the mass-media system's responsibility for forming public opinions comes into focus. It is here a question of conveying as correct a picture as possible of the real state of the welfare society without overtones of sensational nature. In this connection it can be of interest to cite the British Professor of Information Science, Frank H. George. He says the following: "We now step back from the details of data-processing with the intention of looking at its social implications. This involves us in information pollution – the propagation of false information and control pollution – the utilization of information (whether true or false) in order to control the lives of people, with or without their consent" (George, 1977, p. 32). Furthermore, from the same source the following citation is taken: "We should be absolutely clear that in discussing motivational research, and later when we discuss conditioning, brain research and its potential for control, brain washing and all such related topics, we are emphasizing the control aspect of information pollution, and indeed agreed to call such pollution control pollution, since it may deal in either correct or incorrect information, but the danger lies in the use to which the information is put. This may depend on taking information out of its normal context or using whatever information is available to suppress people whether individuals, information of a society or of whole societies" (George, 1977, p. 74).

The danger in this connection is that the mass-media system or the part of it which is connected to the ruling party or parties gives a too rosy view of the state of the social welfare system. On the other hand the mass-media system connected to the political opposition can tend to give a too pessimistic picture of the actual situation. Thus, in the mass-media's information about social welfare, it can sometimes seem that a certain part of it can be classified as information pollution. However, the welfare consumers can, through their representatives in the political decision organs, influence the way of thinking within the welfare political paradigm. So, the forming of a social welfare opinion founded on what is called "objective" facts can be seen in light of what different observers consider to be a desirable social welfare standard and the possibilities of expanding the welfare to new objects. And, therefore, it can be difficult to judge what are objective facts at a certain time – pure facts without filtering and inclinations or, in other words, without information pollution.

Another danger is also that critics of certain abnormal tendencies in the social welfare system can be exposed to some kind of control pollution. This pollution can take the form of citations of speeches or newspaper articles disconnected from their original context. It can also be a question of enlargement of single events and rumors of a rather innocent nature about the critics, but with the aim to throw suspicion on the critics, all in order to give an unfavorable picture of the critics in question. The leading press supporters of the social welfare system representing the ruling political ideas is unanimous in suppressing the critics and his/her thoughts of blasphemy, in other words, they are like a pack of wolves hunting their prey. The leading circles' vested interests of defending the existing welfare system against every presumed enemy have its roots in these circles' tight connections to the individuals who earn their living within the system . The outspoken criticism has to be silenced and if that is not possible it has to be belittled or ridiculed. Nevertheless, as time passes on , the leading circles can bear in mind the criticism or part of it as a basis for certain reforms. However, these reforms then are launched in the name of the leading circles or "nomenclatura".

17.5. Certain Debatable Cases of Social Welfare in Sweden

Now it can be reasonable to refer to the aforementioned Anders Isaksson and his book "Alltid mer aldrig nog" (Always more never enough). The author of this book gives several examples of debatable cases of support within the framework of the Swedish welfare policy among which the following deserve to be selected. In this connection it can be mentioned that the claimants in these cases appealed to higher authorities since their claims had been rejected in the lower instances.

1. The social board in the Swedish town of Lund rejected a request from a father of paying for the purchase of a mountain-bike for his son. The father could not afford to give his son the bike. The superior legal authority's verdict was that the mountain-bike should be included in a reasonable standard of living and the social board had to pay for the bike.

2. A social board in the city of Stockholm rejected a request from the mother of a 19-year-old son where the mother claimed that the board should pay rent for the room her unemployed son used in her home even though the son was living in his old boyhood room. However, the superior legal authority declared that the social board had to pay the questioned rent. The motivation was that the parent's responsibility for upbringing had come to an end and the boy had to be considered as a common tenant.
3. An air-hostess in the city of Stockholm who had given birth to a child requested at the social board to give her 210 hours of child care per month at extra ordinary hours (evenings, holidays and nights) in order to be capable to manage both parenting and her work as air-hostess. The social board declared, however, that she only could get 50 hours per month.

The superior legal authority's verdict was that the air-hostess had the right to get 210 hours of child care per month. But in this case she was not able to pay the whole cost and the social board had to pay the main part of it.

Thus, in this case, it can be presumed that the society shall have the responsibility of standing the costs in order to give the air-hostess the possibility of living the same life as before the child's arrival – flying intercontinental and sometimes being far away from the home several days.

The further comments by Isaksson on this case are the following: "The air-hostess chose to give birth to a child. But that shall not be a cause to change her way of living. Certainly she can choose to change employment or request other tasks with more suitable working hours. However, she does not want to choose these opportunities. So, the society gives assistance, taking over responsibility and costs. The citizen makes demands and the society is delivering demanded services" (Isaksson, pp. 102-115).

In the context of the cases above it can also be of interest to refer to some impressions of the Swedish social policy conveyed by the British researcher, Arthur Gould, according to the following: "In my first visit to Sweden in 1982, the impression gained was that the abuse of social assistance there was a non-issue. Interviews with politicians, officials and social workers suggested that the problem existed but was regarded as very minor indeed.

Social workers are responsible for the administration of social assistance and make it their job to advise on social welfare rights at the same time. There is no attempt to conceal from claimants what right they have to benefits; on the contrary, it would almost seem as if the system operates to ensure that maximum claims are made.

One group of social workers said that they would not dream of providing a client with second-hand furniture – it had to be new. Application forms for social assistance seemed to be short and simple. In short, claimants had a right to social assistance which gave them a good standard of living" (Gould, 1988, p.99).

Relative to the cases above consider the following comments by Isaksson: "And so it goes on, case after case whose final decisions are preceded by large investigations at the social bureau and gathering information which is sent there and back. The social welfare system is living encapsulated in its own world, a glass-cover without mutual reciprocity where the State is playing its role and the citizen his role; the State's role is to give and to make a soft bed, the citizen's role is the right to claim and be given. In this world does nothing so banal exist as a second-hand market for, e.g., prams. There are no friends who stand by with support of things they do not need any longer, and never are there any grandparents who come up with a toy or pajamas for the baby. The officials of the welfare state set norms for all things down to the Consumers Board's approving the cost of the baby's bite ring" (Isaksson, pp. 113-114).

17.6. The Responsibility of Keeping Welfare within the Frame of Financial Resources

The Finnish researcher J.P. Roos writes among others apropos social welfare: "However, it must be emphasized that I do not consider social welfare, nor will I speak of it, as the welfare of the society, assuming society to be a sort of quasi-person. Society consists of the relationship between individuals, nature and the material objects created by the individuals. Social welfare is specifically a concept related to the individual or to a group of individuals. It is possible therefore, to speak of individual social welfare and collective social welfare meaning by this simply that we are interested in either one individual or in a group of individuals" (Roos, 1973, p. 41-42).

But what is the purpose of a social welfare system? The author of this report will try to answer this question according to the following: The basic purpose of a social welfare system is to protect exposed people from starvation, homelessness and other kinds of extreme poverty – in other words, to give help for eliminating distress caused by poverty. But in the gradual development of an industrial society, the demands of more and more efforts from the society to ensure a certain standard for all citizens have arisen. Thus the society has been involved in what can be denoted as a rising social welfare spiral. The economic growth has lead to people's appetite for the society's "welfare cake" to increase and it tends to break the economic frames of the welfare policy. The political decision makers have not had the will nor the courage to resist the vociferous claims from the welfare consumers. And at the same time the "bureaucratic fat" in the welfare administration has tended to be accumulated. What here is said will mean that the welfare generating forces can not at the same time keep pace with the increasing welfare costs, and that can lead to a crisis and perhaps a collapse of the welfare society as described earlier.

An alarming scenario has just been described. This scenario ought not to be regarded as something of a fiction. An evidence of a collapsing welfare state is the case of Uruguay. This country, as known, was one of the world's most developed welfare nations. According to a documentary report "Fallet Uruguay" (the Case of Uruguay) written by two Swedish journalists, Lars Jilmstad and Sven Otto Littorin, Uruguay during the inter-war period was one of the wealthiest countries in the world. But its export mainly consisted of animal products. When the export in the 1950's gradually decreased as a consequence of a hardening foreign competition in the production of animal products, the welfare was running into danger. The existing welfare system, however, was to be prevented and even enlarged by employing more employees in the public sector as a means to combat an increased unemployment. And that resulted in the welfare generating system not being able to keep pace with the continuously increasing costs of the social welfare. The country was during the time after the second World War transformed from a very wealthy nation to something of what can be described as an underdeveloped country, and the country's welfare system was broken down and ended in an economic and political chaos (Jilmstad and Littorin, 1994).

In relation to what has been said in the foregoing, it can be of interest to have a look at the following figure (Figure 47 which shows an interplay between a labor government and trade unions).

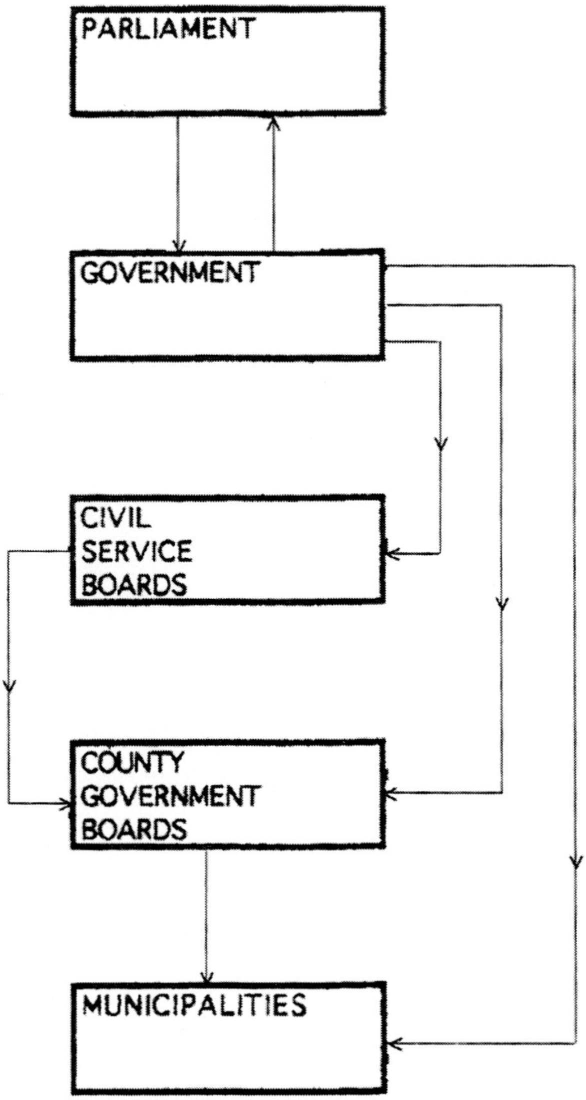

Figure 47. The main features of the Swedish public decision and execution process according to the constitution (from Rasegård, 1993).

So, the figure above is meant to illustrate how the trade unions can exert influence on the social welfare system and, vice versa, how the political ideas of a labor party's program can influence the leading circles of the trade unions . In this connection it seems possible to speak of something like communication vessels. The combination in the figure above can be a danger to democracy. This may result in pressure from the trade unions on the political decision system becoming too strong when it comes to enlarging the social welfare system, without appropriate consideration to the actual financial situation. The "demand machine" will be at top rev and the country risks running into economic and political chaos like Uruguay did.

At last it can be worthwhile to refer to the following citation: "The central questions of human welfare are not solved by models, by bureaucrats, technocrats and scientists, they are solved by men themselves – perhaps with the aid of social science – but through changes in the social consciousness, in the conscious activities of men to improve their lot" (Roos, 1973, p.225).

Chapter 18

SOME DANGEROUS FEATURES IN A WESTERN DEMOCRACY

18.1. SOME WORDS ABOUT A DEMOCRATIC CONSTITUTION AND A META SYSTEM

Previously the welfare system had been described. The designing of a welfare system is as earlier indicated, mainly a political question. Therefore it can be worthwhile to have a look at the main features of the political decision process in a country. The model for this somewhat sketchy study is the Swedish political decisions structure. This overlapping description of the Swedish real power apparatus is based on the British journalist Roland Huntford's book "The New Totalitarians." Despite that Huntford in certain cases has exaggerated it can be of a certain value to look at some of the main features in his description, because they point at certain dangerous tendencies in a democracy which also influence the welfare system.

To begin we can have a look at the principal feature of the Swedish constitution.

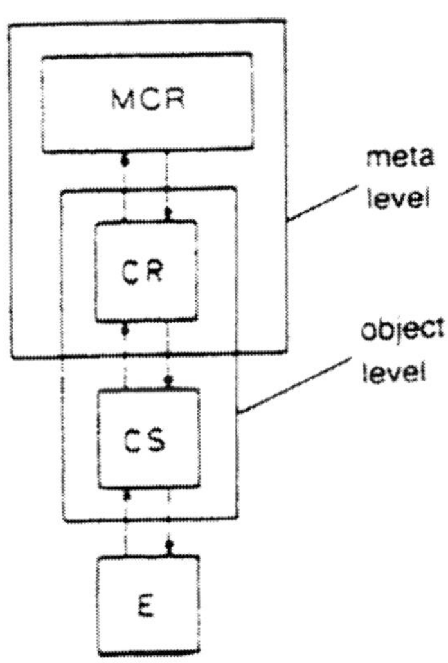

Figure 48. Control and meta control (from van Gigh, 1991, p.390).

As a well-known model of a western democracy (at least a formal one) Sweden can be considered. A schematic of the main features of the decision and execution process according to this country's constitution is shown in Figure 47.

As in all modern democracies, the parliament is the highest deciding organ. The government passes its proposals to the parliament . If not approved, the proposals will be rejected or passed back to the government for re-consideration . The central civil service boards are partly independent decision-making units and partly executing the government's decisions. Within such areas are those where the respective board has the specific right to decide and the government has no right to decide instead of the respective board. This means that so-called ministerial rule is prohibited. This is a unique feature in the Swedish state apparatus. The local administration is divided into county government boards who also are a kind of supervising authorities of the municipalities.

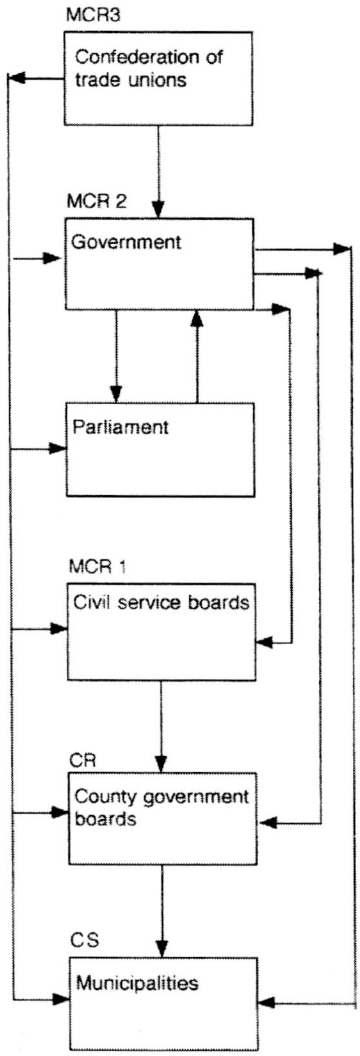

Figure 49. The main features of the Swedish decision and execution process according to Huntford´s view (from Rasegård, 1999).

John van Gigch gives in his book "System design modeling and meta modeling" examples aimed at improving the quality and effectiveness in the public decision and execution process. The most vital concepts in his system view are: metasystem , metasystem paradigm, metacontroller and object level. He defines these concepts according to the following: "Metasystem – a system above the object level system. Metasystem paradigm – a paradigm which models the organization and decision making according to a hierarchy of inquiring systems where different problems are elucidated depending on their levels of abstraction, of logic and of language. Metacontroller – in a control system, the entity exercising control above the controller (CR). Object level - designates the organizational level or the inquiring system below the meta level" (van Gigch , John, 1991, pp. 426-427). A simple figure illustrating the core of the ideas is shown in the next figure.

A motivation for using the meta system ideas according to van Gigch can be summed up in the following citation: "To understand the logic of a system and to be able to exercise control over it, a language and a system of a higher logical order than itself is needed. The validity of this admonition can be exemplified by the inability of two equally strong parties or parties of equal status to resolve a disagreement or a quarrel. Only a decision maker who can survey the issues at a level of discourse above that the dispute will be in position to reconcile them." (van Gigch, John, p. 213)". Van Gigch says further apropos self-regulation in a complex social system: "Self-regulation means autonomy and freedom for members to exercise choice, these objectives may be at odds with any attempts to alter or influence their "appreciative system" in order to obtain consensus." (van Gigch, John, 1991, p. 219).

In connection with what has been said above, it can be reason to ponder the following questions:

a) Who appoints representatives ?
b) The criteria after which the representatives is appointed?
c) Are the appointed representatives also represented on other levels?
d) Is it possible for the representatives to act uncommitted or which commitments can be found?
e) Do the appointed representatives have competence enough?
f) Which is the type of image paradigm, is it ruling, collective or individual views?

In connection to point e) the following anecdote can be cited concerning appointment of a head of a social department. The question from a reporter to a politician responsible for the appointment: "Has the appointed the necessary competence?" The politician answered: "Do you question his competence? He has enough competence because he has been a member of our party for many years."

18.2. A Corporate "Democracy"

After these short descriptions of the Swedish constitutional state apparatus and certain essential features of van Gigch's metasystem ideas it can be of interest to refer to Roland Huntford's book "The New Totalitarians" where he gives another picture of the Swedish state apparatus than exhibited in Figure 48. In his book he is pointing out the Swedish society as a form of corporate state disguised in a costume of formal democracy. Huntford's view of the public decision and execution process can be summed up by the following schematic.

(It ought to be noted that the Swedish king only has a pure representative function without influence on political matters.)

The confederation of trade unions makes proposals to the government. The government passes the proposals to the parliament which makes decisions on behalf of trade unions. Hereby the parliament puts the proposals in the form of laws and ordinances.

The confederation of trade unions can through their ombudsmen and other politically active persons of their organizations exercise control of parliament, civil service boards, county government boards, decision assemblies of municipalities and university boards. It ought to be noted that this control is exercised in the name of democracy and based on common political elections.

The indoctrination of the people into an anthill mentality conformed to the ruling party then begins at the day care centers and at children's departments of libraries and continues through the school system.

Applying van Gigch's ideas of metasystem to Figure 49, it can be interpreted that no less than three metacontrollers can be seen. However, it ought to be observed that the parliament can not in this connection be considered as a metacontroller because it only has the function of a "transport company" for transporting the government's proposals into legal frames. Because the metacontrollers also in several cases are the same persons both on higher or lower level, it will be a danger of conformity to such a high degree that it can be compared with that of a totalitarian system. It has been a kind of "short circuit" in the system due to the fact that the politicians tied to the trade unions and the leading political party wear too many "hats" at the same time, and this increases the danger of an ant-hill conformity.

Even if it is possible to admire the society of an ant-hill, its structure and functions, seen from a pure biological viewpoint and as a component of a larger biological pattern, transformed into a human society it would be a terrible example – an example that thinking individuals must keep their distance from if their unique identity shall be preserved. The comparison with the anthill, however, is not complete without mentioning the intellectual consciousness which is the characteristic feature of human beings. This intellectual consciousness means among others that the individual is conscious of his life-span and is able to form symbols in order to give durable expressions of his thoughts – thoughts which can form imaginations that anything can continue even when the earthly life-span has come to an end . In this lies the greatness of man. The human gift of forming symbols has developed the communication between human beings. By using written symbols man has possibilities to further his development in a completely different direction than he would have been able without symbols of oral language. The individuals of an anthill certainly can communicate with each other according to inherited patterns . But these patterns equal only elementary conditions (robotic expressions) in order to make it possible for the anthill to survive.

Going back to Huntford's book, this meaning is to show that Sweden has a type of "nomenclatura" formed by the ruling party and caused by this party's long time in power. There is a special danger in a political party's dominating role for so long a time in that its voters can be trapped in a myth that only the ruling party is capable of taking the social responsibility for the citizens. The ruling party has formed a paradigm which is very difficult to break through . A special danger here is that the people can show less interest in political matters which favor the clique of those politically interested who will form the policy. If this limited circle consists of people trapped in a political paradigm, which in turn is not up-to-date with the real situation, the risk will be that the decisions will not adequately cope with reality.

It must be noted, however, that the political representatives ultimately depend on their party apparatus through which they are nominated. If the party apparatus similarly is tied up in a paradigm of rigid dogmatic rules, the danger again is that the decisions can be dictated by dogmatic views instead of pragmatic considerations. Another danger is that information which is not suited for a certain paradigm can be suppressed. Certain individuals in key-positions can find this information so inconvenient that they "put the lid on." The individual who despite that, insists that the information should be considered risks being mobbed or replaced to a position with lesser insight. This can be a serious problem both in public and private organizations as well as in the community of science. An indication of that is the formation of the Swedish organization Exodus in 1983. This organization was formed by officials who believed that the fact that they had attempted to express their opinions or to convey certain inconvenient information had resulted in replacement or other problems in their workplaces . Their opinions were, in other words, unfit for the ruling paradigm.

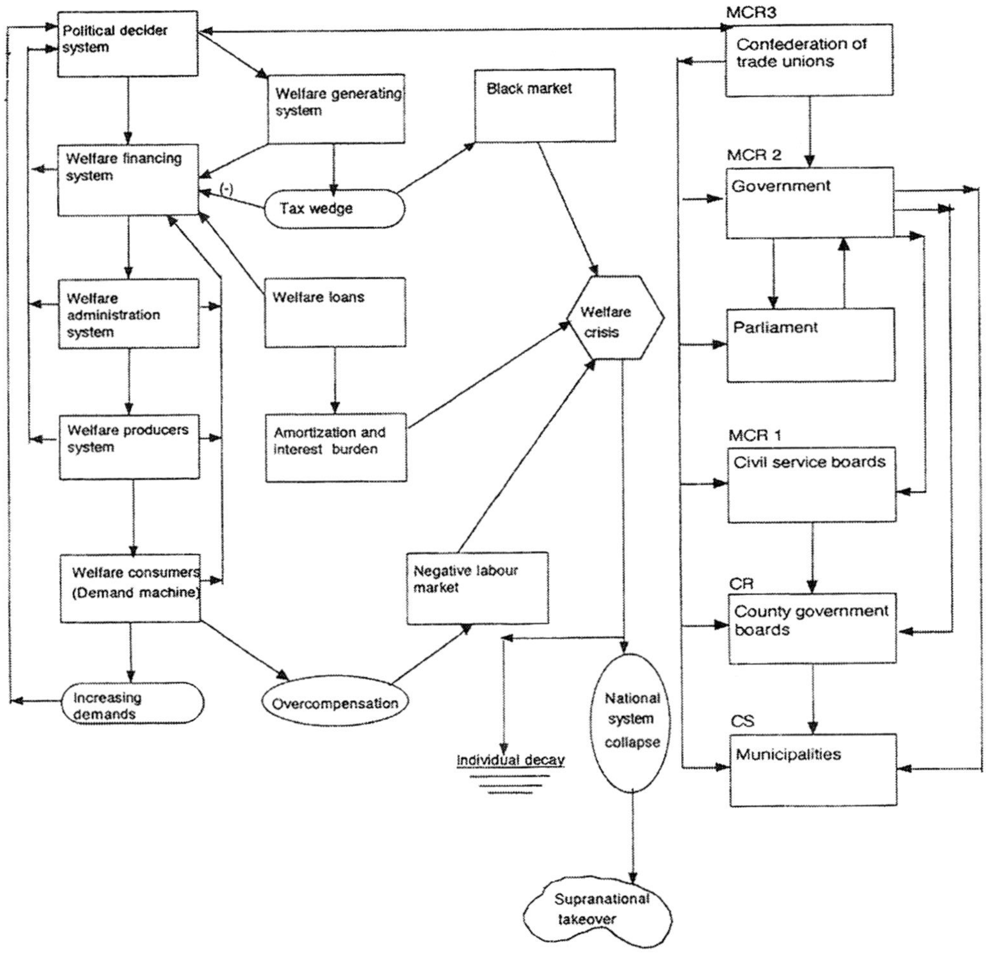

Figure 50. The Social welfare influenced by a corporate thinking (from Rasegård, 1996).

At last it can be of interest to look at how the welfare system can be coupled to a corporate democracy. So the figure above is meant to illustrate how the trade unions can exert influence on the social welfare system and vice versa, how the political ideas of a labor party´s program can influence the leading circles of the trade unions. In this connection it seems possible to speak of something like communication vessels. The combination in Figure 50 can be a danger to democracy. Pressure from the trade unions on the political decision system can be too strong to enlarge the social welfare system, without appropriate consideration of the actual financial situation. The "demand machine" will be at the top and the country risks running into economic and political chaos like Uruguay.

However even in Huntford´s view of the Swedish state apparatus can seem to show certain features of something like an apparatus in a totalitarian state; the important difference is that Sweden has free elections of its political representatives. Thus at last the voters have the ultimate responsibility of forming a democracy free from the not so good features of corporate organizations' influence on the state´s governing matters.

CONCLUDING COMMENTS

The author of this book has made an attempt to show how systems of different kinds influence the living of man from cradle to grave. However, the descriptions of systems can only be regarded as an overview of the jumble of systems and paradigms combined with social conventions which are influencing man´s pattern of living. Maybe, some of the first part of this book can seem tedious for a reader not interested in organizational systems. However, these systems can be regarded as a base for development of the systems on a higher level, for example, the welfare system.

Concerning the term man-made system, it can be noted that this denotation has been used as a denotation of organizational systems made by organizational experts. But it is not a welfare system created by experts. Yes it is true, but it is a special kind of man-made systems even if the detailed structure is a work of experts. However, the overlapping structure is a work made by politicians and built on political decisions. Thus, a welfare system can be regarded as an overlapping national system. In this book the welfare system has a rather large space. For which reason? The answer is that a welfare system can be said to form a base for the development of the society. Without a functioning welfare system it is at risk of malfunctions of the societal apparatus. The manifestations of these malfunctions are unsatisfied citizens together with trade unions who monopolize the welfare system for certain political aims . That can imply that certain democratic functions can be tricky.

A very important component in forming a welfare society is that scientific research can get enough financial resources and that it is possible to do such research free from unnecessary governmental restrictions. However, it is important in this context to make a distinction between real science and pseudo science. In this connection it can be worthwhile to forewarn for such orthodox paradigms that ban certain research of anomalies out of the established paradigms. A typical example of such research is parapsychology which to certain scientists is considered a pseudo science. Thus, it can be noted that research in anomalies often is regarded as something awful like the cat has brought indoors . Despite this fact it can be desirable that researchers of certain anomalies can be met with a more generous attitude from scientists working within established paradigms and that such a type of research will not in every case be regarded as superstition or fraud.

ADDITION

THE FAIRY TALE OF INTRODUCING A NEW ECONOMIC MANAGEMENT SYSTEM IN MIDDLETOWN

ABSTRACT

This article can seem provocative but it deals with problems that sometimes are neglected. The article can be seen as a kind of satire which reflects clashes between different opinions which were aroused in connection with the decision to implement a new system of economic management (budgeting, accounting and data processing). In the article are described something of the intrigues connected to decisions on how to approach the problems of introducing a new system. The story is not fiction, but rather is grounded in reality; it shows examples of two different outlooks related to aspects of how an investigation process should be done according to "model up" or "model down."

Key Words: critical subsystems, financial failure, personal policy, systems designing, systems thinking.

Once upon a time there were some reformers who wanted to introduce a new system of economic management in a little town which by the common people was named Middletown. However, its real name was quite another. This little fairytale will deal with the toilsome work of investigations connected to introducing a proposed system. That it is a fairytale does not necessarily mean that what here is told could have happened in reality. As known reality can sometimes overthrow fiction.

The place for the story thus is little Middletown situated at an idyllic bay of an archipelago somewhere in Sweden. Middletown´s history is rather old and goes back to the Middle Ages. The town was according to a tale founded by a cousin of King Erik the lisping and limping who in despair of his prank later became a monk, and now in his function as the town´s patron saint is standing a statue in the central square. About how the name of Middletown has originated the scholars dispute. Some consider that the name had arisen because the inhabitants of the town always had tried keeping the golden middle-way in politics – not too much on the left and not too much on the right. Others believe that the name has its roots in the people´s well-documented economical handling of money – by not overspending. A little story that gives a glimpse of that has been told. It happened once upon a time that an old money-lender should be buried. Then some of his clients became remorseful for not being able to pay the interest in time and showed that by throwing some coins in the grave. Among the present clients was a man from Middletown. He then showed his remorsefulness by throwing a cheque down in the grave to his old creditor. About the political majority in this little town can be said that it can shift from case to case depending on the fact that certain politicians have some difficulties seeing the difference between right and left. This situation can be similar to a night-orientation where certain competitors lost both lamp and compass. However, at last the affair usually will be settled according to the implicit instinct – not too far to the right and not too far to the left. What here has been said can, however, cause a certain prolonging of the decision process. But now over to the true basis of the story.

The reason for the action by the hearty reformers, as mentioned above, was that the agencies which these reformers represented considered that the present system was old-fashioned and as soon as possible ought to be replaced with a new one. They were seeking a system where the heads of the town´s agencies, whenever they wanted, could press a button on a computer and get the actual economic state on the screen. On this screen one should even be able to see the effects of different economic decisions and thereby provide the politicians with rational foundations for decisions.

The investigation began with the formation of a committee. This committee was born through a virgin birth, that is, the members of the committee appointed themselves without any formal decision from any authority. After about a year of pondering and meetings, the committee was ready for a preliminary evaluation which concluded that the town be recommended to introduce a new system for accounting and data processing. The committee also recommended that the Town Government appoint an unbiased investigation organ for investigation of the whole economic management system. The Town Government decided to accept the recommendation and appointed an investigation group which, aside from experts from the employee area, also included heads of the town´s principal agencies. This investigation group had to report to a reference group of politicians.

Now the question arose who would be leader of the new investigation group . Then it suddenly appeared that the town´s political bigwig, the mayor Napoleon Alexander, gladly would see that a head of one of the town´s most important agencies, the agency of Finance and Accounting, the Director Charles Sacking, was displaced. Then the mayor had the opportunity to displace this, in his eyes, too independent of a person with one who was more obedient without so many opinions of his own. The mentioned Sacking was persuaded, after some pressure, to resign from his appointment as an agency head and instead become the leader of the investigation group, which had to prepare for and introduce a new system of economic management. One of the first points on the agenda would be to investigate the budget system.

After a time, however "it was a fly in the ointment" when the dictatorial mayor suddenly went into action. He now considered that the time was ripe to get rid of the man who so many times had made him irritated. And he succeeded with the help of obedient assistants within the official cadre to find a reason. He called upon the leader of the investigation group and ordered him in a brusque manner to make changes in part of the investigation before it was presented for the political reference group. Sacking objected that he was appointed in order to make an unbiased investigation and the mayor "the great little man," who was rather short in growth, answered with "here, I am determining." After this talk the time had come for Sacking to give in his application of resignation, because he considered that he did not have any possibility to continue his work under this form of dictatorial governing. The secretary of the investigation group, Fabian Middleman, before then had in vain tried to persuade him that despite Alexander's behavior, to present the partial investigation for the political reference group.

Now our mayor at last had reached his goal and he gladly recommended Sacking's dismissal to the City government. However, the situation became a bit confusing. The question was now how the measures against Charles Sacking would be justified in a report to the political opposition. This defence report then pointed out that the program of investigation which Sacking had drawn up was not relevant and must be an object of revision. Sacking had not been competent enough for his work. Consequently, in order to avoid delaying the urgent matter, it was necessary to employ another investigator. Alexander then proposed that a female economist, Susanne Muddling, be appointed. She was, according to him, a very bright star who soon would have a new program ready with required priorities among the various investigation matters. That these priorities already were done by Sacking and his group was not to be mentioned. Here the question was that at any price sweep out the traces of Charles Sacking's "miserable" work.

After Susanne Muddling in her own way had finished her investigation of the earlier investigation and thereafter had disappeared out in the periphery a deal of negotiation and intrigues began before the town's wise men at last decided to appoint a new investigation group. The goal-setting for this new investigation , however, was more narrowed than the first investigation. The new investigation group should, namely, on the first hand devote its time to design a new accounting system and, if it would be necessary, to propose a new data processing system.

Thus, the time had come to start another investigation with new, fresh ideas. Young people without much experience but with good education were employed. These youngsters, glowing with enthusiasm, began their work under the guidance of a newly employed consultant, Solomon Wiseman. The goal of the new investigation group, as the directive from the mayor ordered, was to develop within two years a new system of accounting and data processing and hereby solve the Gordian knot which Charles Sacking, according to his slanderer, had failed to solve. By permission of the Mayor Napoleon Alexander, Wiseman decided to begin to investigate the data processing system. Four years had now elapsed since the virgin birth of the preliminary investigation group. Nothing of importance for the economic management system had during this time been done with the exception of an investigation of a peripheral nature.

However, after a time another problem began to emerge. This problem was connected to the town's center of computer-based data processing, called the data central. One of the town's leading politicians, namely, had made a statement that it was suitable to break out the mentioned organizational unit from the agency of Finance and Accounting. Based on his statement there were two different opinions aroused – one which considered that the organizational structure should not be changed and another which declared that the most suitable option should be that the unit form a separate agency. The reason for this was that the data central aside from working up and arranging accounting matters from the agency of Finance and Accounting was also working up and arranging technical and statistical data from different agencies of the town. Both opinions were lead by two of the town´s principal officials , Goliath Battleman respective David Armstrong . The first mentioned was the leader of the host who was against a change while the last mentioned considered that the agency of Finance and Accounting had always held a "dead hand" over the development of data processing and therefore wanted to "liberate" the data central from its "oppressor." Since the matter several times had been postponed at the meetings of the town government, it was settled at last by having the data central become an agency of its own. This decision had been preceded by numerous lobbying by the representatives of both parties.

When this little fairytale ends there is not yet any concrete proposal for a new system of economic management despite that nearly six years that had elapsed since the first official group "cut the first sod for" the town´s investigation of the mentioned system. The numerous meetings, namely, had mostly been discussions about which new system of data processing would be suitable for the town. An information analysis had never been done which could show which types of information a new data processing system would produce or how the raw data material that is worked up swould be structured. The teller of this fairytale can like other tellers of fairytales finish his tale by saying that if the investigation had not slumbered in then it is continuing its little life on its own. Yet, maybe, the taxpayers hope that the new system of economic management would give improved foundations for economic decisions and hereby make it possible to spend their money in a more rational way. Like some other fairy-tales even this tale should give birth to reflection. Which are the problems that can be distinguished from the story material? Below attempts have been made to answer this question. The main problems can be summarized according to the following:

1. Anchoring the investigation to the political decide-function.
2. The way of approaching the system problems.
3. The need of a wholeness view.
4. The need to clarify the system as an integrated wholeness .
5. The need to clarify where one will land.
6. The way of displacing "unsuitable persons" to investigation missions or other missions out of the ordinary work of an organization.

1. *The anchoring to the political decide-function* in the beginning was good – in pure theoretical meaning. However, the wholehearted engagement of the politicians had been missing. This phenomenon can partly be due to the "great" town politician Napoleon Alexander´s filtration of the investigation information.
2. *The way of approaching the system problem* had from the beginning a good foundation, when one understood that one first ought to determine how the budget had to be designed. That was necessary if the economic management system as a whole could in the best way serve as the politicians´ foundation of decisions. Second, the demands should be investigated which the budget had on the accounting and data processing systems. It can here be mentioned that the municipal economy, in contrast to what is valid for a private enterprise, by the municipal law in Sweden is completely tied to the budget.

3. *The need of the wholeness view* has not been sufficiently considered when the Wiseman's investigation group began to work. An extract from "Conceptual Considerations" can serve to illustrate the importance that the wholeness view is not being neglected.

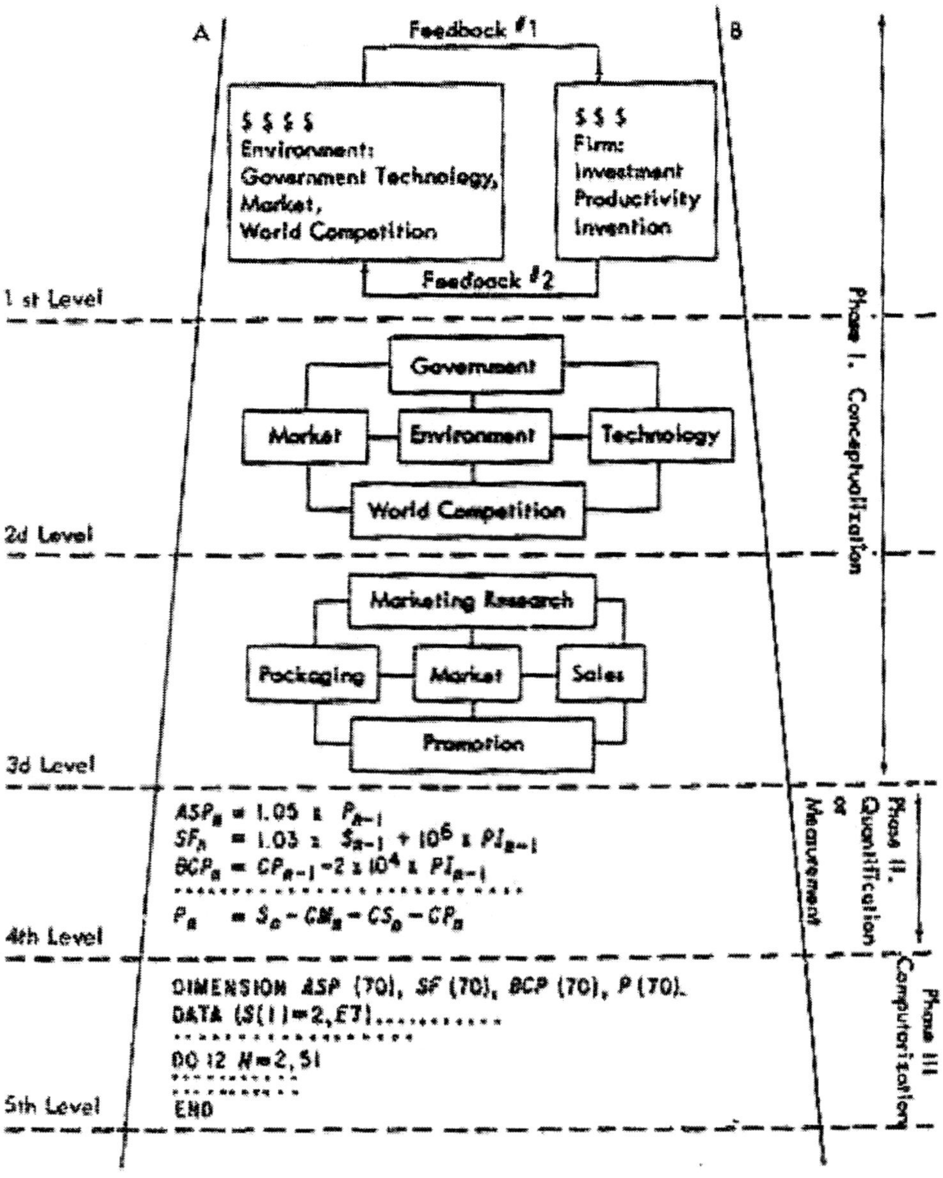

Figure A.1. Modeling of the environment-organization interaction system through the cones of resolution (Kefalas and Schoderbek, p.256).

Figure 1 above (Figure 9-6 in " Conceptual Considerations") gives a picture, despite it being incomplete, can be applied to a municipal activity. However, certain information about the problem complex can be given if one in respect of Middletown considers that the levels of the figure can be translated according to the following:

Both first levels represent the environment (state authorities, citizens and associations of different interests). The third level can be assumed to represent the budget system, the fourth level the accounting system and the fifth level, at last, the data processing system. The extract below from "Conceptual Considerations" speaks for itself.

"To recapitulate it is imperative that the systems scientist proceed from conceptualization to computerization and not vice versa, as is indicated by the upside down pyramid in Figure 9-6. A better grasp of the problem along the lines of Phase I will enable the researcher to develop a better measurement method and a better computer program rather than force the problem into a preconceived computer program.

As can be seen from Figure 9-6, the systems scientist initially loses quite a bit of the real phenomenon of the environment-organization interaction system because of the abstract and aggregative nature of the manager´s conceptual model. These losses, however, are rather moderate when compared to those that would be incurred were the manager to follow the "model-up" method rather than the "model-down" approach which is advocated here. The conventional modeler will begin at the base of the cone of resolution and then will proceed upward.

The typical analytic thinker´s cone of resolution can be imagined as being the mirror image of the systems scientist´s cone of resolution. A quaint man´s model will most likely begin at the fifth level of the cone of resolution depicted in Figure 9-6. In working his way up (model-up) he will incur certain losses attributable to his narrow and precise quantification of certain relationships of the organization-environment interaction system. However, unlike the outcome for the systems scientist, his losses tend to increase as he ascends his cone of resolution until he crosses the boundary of reality (as perceived by the manager) . From there on he loses touch with reality, eventually reaching a point of maximum irrelevance by pursuing certain solutions to problems which he alone can understand and interpret.

It is unfortunate that some, parading under the banner of operations research and management science, use their techniques as procrustean beds upon which managerial problems are amputated and distorted so that they exactly fit the sacred box of the quant man." (Schoderbek, Kefalas and Schoderbek, 1975, p. 260).

It can be added that what has been said above has some resemblance with the tale of the Indian rope-trick where the fakir when climbing upward sees his the rope suddenly disappear into a misty nothingness.

1. *The need of creating clarity of what shall be involved in the system* is conspicuous by its absence. In the Wiseman investigation the goal it narrowed down to designing a data processing system without consideration of the fact that this system is only a part of the economic management system. The connections between the main parts, namely, can be indicated as follows: The budget is an activity program and thereby declares the act of political will. Therefore the accounting system can be regarded as a subordinate device of the budget for systematic registering of transactions connected to budgeting activities. The task of the data system in this connection is that, on the basis of the codes in the accounting system and other given conditions, it provide the municipal agencies with economic reports.

2. *The need of clarifying what one will achieve – which goal-setting one has –* seems in some way to have been lost in the handling. By starting with the presumption of investigation a new system of data processing without a foregoing information analysis, e.g., what types of information the new system is expected to provide and to what it shall be used, one risks neglecting important demands on the system which can arise during the implementation phase, for example, demand from the politicians concerning changed budget conditions.
3. *The way to displace "unsuitable" persons from investigation missions* can be regarded as a questionable method which, however, seems to be applied both within public administration, academic organizations and private enterprises. The question here is if the displaced person regards the new tasks meaningful enough in order for him/her to be effective in the new role. In the tale here told, Charles Sacking is a powerful person with intentions to continue with his work when he is abruptly forced to accept a displacement. However, other examples can be found where the "investigator" has lost interest and is working in a sloppy way. This can be a serious problem which has to be tackled in a proper way, but unfortunately often is neglected. An indication of that is the formation of the Swedish organization Exodus in 1983. This organization was formed by officials who believed the fact that they had attempted to express their opinions or convey certain information had resulted in replacements or other problems in their workplaces. These opinions or this information was, in other words, not fitted for the "paradigm."

SUMMARY

The described scheming situation in the fairy-tale of Middletown can not be considered as unique in public activities. A certain tension can always arise between politicians and officials. In this connection it can be mentioned that an official in public service according to Swedish regulations hardly can be dismissed if he is not guilty of mishandling his position. In order to get rid of him, the politicians must transfer him to another position. Sometimes he, as a gesture of good will from the politicians, can get some economic compensation which often has been denoted as a "side-jump" upwards. This type of compensation has also sometimes been named "unfitness add". In the case described, the displaced was recognized as a very competent official who, however, had the misfortune of touching a mighty politician´s "ailing toe."

In the new investigation that had been started after Charles Sacking had been dismissed, one other view of the problems was presented. It was considered that it was not necessary to take in the whole system of economic management in the investigation work. Instead, one could begin to seek a system of data processing which was large enough to handle all thinkable problems in economic management. The other parts of the economic management system seemed to be considered as something of a black box. That might sound very good. Maybe, the figures in this fairytale can succeed and in this way provide Middletown with a new effective economic management system. Perhaps any good fairy appears swinging its magic wand over the investigation.

After reading this story with its different views of investigations and interests as the tale has given glimpses of maybe the reader, like Pontius Pilatus, asks: "What is true?" The teller of this saga would then answer: "The truth is relative." Perhaps one like the old satirist Bernhard Shaw can say: "The golden rule is that there are no golden rules."

Perhaps the reader now asks where Charles Sacking was going – but that is as Kipling said another story.

At last a little explanation. In the saga one can get the impression that Middletown is a little town. However, that does not necessarily mean that it must be a question of a little town. Even if the town according to global measures can be regarded as not so big. As known everything is relative and according to Swedish measures the town can be considered to be on the larger side.

EPILOGUE

Several years have now elapsed since the time this saga had ended and Middletown has at last achieved an acceptable system of economic management. The costs have been high and the failures have been many. However, suddenly the Town government learned of happy things. The City of Stockholm, capital of Sweden, namely had begun to speculate on the stock exchanges and that seemed to be a very good affair for this town. Now certain wise men of the little town considered that they must grasp the golden chance like "Big Brother" has done. Here appeared possibilities to consolidate the financial position of the town. In order to give the speculations a speedy start an assistant to the town – cashier was sent to a one-week course in stock managing. After he had attended this course he proposed that a data processing system specially designed for stock affairs should be purchased. However, the "great municipal man" Napoleon Alexander believed that the town could not afford to purchase such a system since the costs of the economic management had been abnormally high. So, he ordered the earlier mentioned assistant , who was a clever young man, to compose a system of his own. The assistant did so and named his system "Symphony." In the beginning all the affairs of the stock market went very well and the assistant made rather high profits by his speculations. All was well. The money flooded into the town and the assistant was recommended for promotion. However, suddenly the "Symphony" ended abruptly. The system broke down. And the poor assistant became confused. He was not able to read out the actual statements of his affairs. Heaps of transactions could not be registered . The town auditors began to scrutinize the affairs with help of external consultants. When the investigation was done it was evident that the town had lost a considerable sum of money. The great man, Napoleon Alexander, took the consequences and gave in his application of resignation. The town treasurer and his assistant immediately were secluded from their work. The town auditors produced a charge to the three men who were involved . They were charged with faithlessness to the head principal (the town) and the question of imprisonment hung in the air. However, the great man, Napoleon Alexander, had employed clever lawyers for defense. And the whole story ended with all persons involved being acquitted. The defense, namely, had stressed the fact that the involved persons had worked with no interest to enrich themselves but only to enforce the town's financial position. Napoleon Alexander was rendered a rather high pension and the two others involved were transferred to other appointments in the municipal administration.

Now over to the financial failure in the City of Stockholm which happened in 1986 and was described in the international press. The description is here based on official protocols and documents completed by interviews with officials in the City Hall of Stockholm. The main difference in relation to Middletown, however, is that the leading politicians denied all knowledge of the speculation on the stock exchanges. In everything else the pattern of events was the same. Like Middletown Stockholm used a "home-made" data processing system for the speculation on the stock exchanges which broke down abruptly creating chaos in the registration of transactions. Like Middletown those involved were acquitted with the same reasons as in Middletown. The foremost scapegoat, the head of the agency of Finance and Accounting who gave in his application of resignation was rendered a rather high pension. The others involved were transferred to other positions. It can be mentioned here that the financial loss for the municipality was 444 million Swedish crowns. Below a description and analysis have been done according to Miller's 20 critical subsystems. These subsystems have been described earlier in this book. The description of the case and analysis below is an extract from an article in the journal "Behavioral Science" written by the author of this paper.

A financial failure is seen in the perspective of Miller's system design.

An analytical diagram with comments which depict the failure according to Miller's critical subsystems follows below.

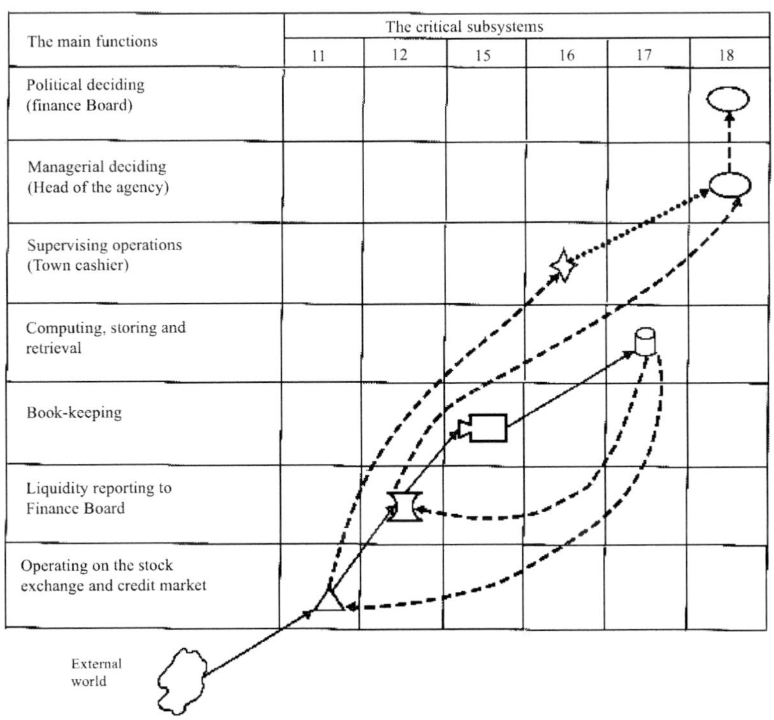

Figure A.2 The finance and liquidity management (from Rasegård, Behavioral Science, 1991, vol. 36, p.96)

First it must be stated which function or position the mentioned municipal operator on the credit market and stock exchanges would have in an organization built on the principles of the critical subsystems. It can then be concluded that this function is connected to the *Input transducer* (11). This conclusion is founded on the following statement from *Living Systems* (Miller 1978): "Information in the form of money and credit enter societies from outside in the form of payments for goods, loans and sometimes subsidies. Most of these exchanges are in the form of credit , which is transduced in the same way as other symbols on paper (Miller, 1978, p 784)" This can also be applied to a municipal organization.

The official at the Cash and Loan Department who performed the operations on the credit market and stock exchanges was an official who, without sufficient knowledge of municipal law, was given the responsibility of purchasing and selling assets and who had to finance his operations with short-term loans on the credit market. He had to seek advice from external sources because no other official within the agency had the required knowledge of the transactions in question . The associator and the decider functions within this special part of the cash and liquidity management were out of function. Then where should both of these subsystems be found in the normal case?

The answer is that when the cash and liquidity handling is seen as a separate system, the decider function is to be managed by the head of the agency of Finance and Accounting, while the associator function is managed by the Town-cashier. The internal transducer function is found within the Control Department and is managed by the official who works out the liquidity reports to the Finance Board.

The decoder function is found at the Accounting Bureau within the Accounting Department and is managed by the official who is responsible for ensuring that the accounting codes on the verifications are correct, and who works out the book-balance of the financial administration as a whole. The decoder function also includes the official within the Accounting Bureau who has the routine task of formal control and of preparing the materials for data processing at the municipal computer center. In this case this center has the function of memory, while the function of channel and net contains the computer net, the telephone exchange in City Hall, and City Hall´s manual message unit. Other subsystems, on which this system stratification has to rely, are found at various organizational levels of this system´s real activity sphere. The system stratification in question is in Figure 8 which shows the key subsystems in this special activity.

The hierarchical organizational structure within the Agency of Finance and Accounting has led to a kind of "preserve thinking" where each manager of a department or bureau attempts to defend his special sector of the total activity. And this situation creates an atmosphere of rivalry. That was one of the causes to the failure, as has been stated by the Town-auditor. As can be seen in Figure 8 the flows of information are marked ______________ and _ _ _ _ _ _ _ _ _and .... The solid line indicates a continuous flow of information from *Input transducer* (11) (the operator) to *Internal transducer* (12) (the official working out the liquidity reports for the Finance Board), and *Decoder* (15) (the officials who carry out the tasks connected to bookkeeping. That means that the basic bookkeeping passes to *Memory* (17) (the municipal computer center) continuously . So far all is well. But what structure is there? As the Town auditor has pointed out, verifications in certain cases were missing. In other words, there was nobody who saw that the verifications had the right accounting codes, that the amount was right, and that the transaction in itself was done within the frame of legislation and regulations. So, the solid line indicates a continuous flow of information of different qualities. The functions of *Internal transducer* (12) and *Decoder* (15) convey information without hesitation and that thus is possible to say that these subsystems are malfunctioning in respect to their duties within the system stratification. The solid line from the outer world to *Input transducer* (11) indicates the information which forms the base for the operators acting on the credit market and the stock exchange.

The results of the operator´s activities are registered in *Memory* (17) which contains an amount of different liquidity assets. *Memory* (17) has in this respect the same function as inventory bookkeeping. It gives an account of different types of liqudity assets at an actual time. But the system of data processing at sporadic times malfunctioned. That means that feedback to *Input transducer* (11) at certain times was broken, as well as to the *Internal transducer* (12). This is indicated by the broken line. *Input transducer's* (11) contacts with *Associator* (16) (the Town-chassier) was not as good as it ought to be, which the broken line indicates. And the connection between *Associator* (16) to *Decider* (18) was almost out of function, which the dotted line indicates. The reports from *Internal transducer* (12) to the managerial decider function (handled by the Head of the agency of Finance and Accounting which was passed further to the political decider function (the Finance Board), was incomplete because they were not reported in an analytical way" (Rasegård, 1991, p. 94-97).

Sic transit gloria mundi.

REFERENCES

Ackoff, R.L (1984). *A guide for controlling your corporation's future*. New York: Wiley.

Arbib, Michael A. (1977). *Computers and the cybernetic society*. New York: Academic Press.

Ahari, Parviz (1992). *Living systems theory and break-through thinking: Purpose oriented living systems analysis*. Report presented at the 36th annual meeting of the Internal Society for Systems Science, Denver, CO.

Barrow, John D. (1991). *Theories of everything – the quest for ultimate explanation*. New York: Oxford University Press.

Bauer, Henry (2000). *Editorial. Journal of Scientific Exploration*, Vol. 14.

Beer, Stafford (1972). *Brain of the firm*. London: Allen Lane/Penguin.

Bertalanffy, Ludwig von (1968). *General Systems Theory*. New York: Braziller.

Brown, Hanbury (1986). *The wisdom of science*. London: Cambridge University Press.

Checkland, Peter (1991). *Systems thinking, systems practice*: Wiley, New York, China Daily, February 8, 1991.

Cho, S. And Runeby, N. (1988). Traditional thought and ideological change , Sweden and Japan in the age of industrialization. Stockholm: Japanogical studies and Department of Japanese and Korean, Stockholm University.

Clemson, Barry (1984). *Cybernetics , a new management tool*. Tunbridge: Abacus.

Davis, Paul (1987). *The cosmic blueprint*. London: Heineman.

George, Frank (1977). *Machine takeover*. Oxford: Pergamon Press.

Delbrück, Max (1986). *Mind from matter* (posthumous edition): Blackwell Scientific Publications, London.

Gigch, John van (1991). *System design, modeling and metamodeling*. New York: Plenum Press.

Goleman, Daniel (1995). *Emotional Intelligence*. Bantam Books.

Gould, Arthur (1988). *Conflict and control in welfare policy – the Swedish experience*. London and New York: Longman.

Greene de, Kenyon B. (1991). Rigidity and fragility of large sociotechnical systems' advanced information technology – the dominant coalition and paradigm shift at the end of the 20th century: *Behavioral Science*, vol. 36.

Greyson, Bruce (2001), Introduction of Tim Dinsdale winner: *Journal of Scientific Exploration*, vol. 15.

Hilgard, E.R. (1948). *Theories of learning*. New York: Appleton-Century-Crofts.

Huntford, Roland (1971). *The new totalitarians*. London: Penguin Press.

Isaksson, Anders (1994). Alltid mer, aldrig nog: T. Fisher and Co, Stockholm.

Isaksson, Anders (1994). *Alltid mer*, aldrig nog: T. Fisher and Co, Stockholm.

Jahn, Robert (2001). 20th and 21st century science: Reflections and Projections: *Journal of Scientific Explorations*, vol. 15.

Jilmstad, Lars and Littorin, Sven Otto (1994). Fallet Uruguay (the case of Uruguay). Stockholm: Timbro.

Johnsson L. And Tomlinson, C.M. (1994). A critical review of the fundamentals of knowledge engineering: *Systems Research and Information Science*, vol. 6.

Kafatos, Menas and Nadeau, Robert (1990). *The conscious universe – part and whole in modern physical theory*. Springer-Verlag; New York.

Klir, George J. And Folger, Tina A. (1988). *Fuzzy sets, uncertainty and information*. New Jersey: Prentice Hall.

Kocken, Manfred (1975). *Information for action*. New York: McGraw Hill.

Kruszelnicki, Tommy. *Systemologi,* 2001, vol. 3: Fimas, Haninge, Sweden.

Lipsey, Richard G. And Steiner, Peter O. (1978). *Economics:* New York, Harper.

Lovelock J. (1979). *Gaia - a new view of life on Earth*. Oxford: Oxford University Press.

Mason, Richard O. and Swanson Burton E. (1981). *Measurement for management decision*. Reading, Massachusetts: Addison- Wesley.

Nadler , Gerald and Hibino, Shozo (1991). *Break-trough thinking*. Prima Publishing and Communications.

Mainzer, Klaus (1994). *Thinking in complexity – the complex dynamics of matter, mind and mankind.* Berlin and Heidelberg: Springer-Verlag.

Miller , James Grier (1978). *Living systems*. New York: McGraw Hill.

Miller, James Grier and Jessie L. Miller. Introduction : The nature of living systems. *Behavioral science*, 1990, vol. 35.

Nalimov, V.V. (1981). *Faces of science*. Philadelphia : Isi Press.

Popper, Karl and Eccles, John C. (1981). *The self and its brain*. New York : Springer Verlag.

Popper, Karl (1994). *Knowledge and the body-mind problem* (edited by M.A. Notturno. London and New York: Routledge.

Pylkänen, Paavo (1989). *The search for meaning – the new spirit in science and philosophy*. Wellingborough: Crucible.

Pyhrr, Peter A. (1973). *Zero-base budgeting*. New York: Wiley.

Rand, Ayin (1957). *Atlas shrugged*. New York: Random House.

Rasegård, Sven (1986). Using ideas of cybernetics in organization development. *Computers, Environments and Urban Systems,* vol. 10.

Rasegård, Sven (1991). A comparative study of Beer´s and Miller´s systems designs as tools when analyzing the structure of a municipal organization, *Behavioral Science*, vol. 36, pp. 83-99.

Rasegård, Sven (1992). *Informatics and systems science applications for analysis and structural improvement of a municipal organization*. Stockholm: Stockholm University and Royal Institute of Technology.

Rasegård, Sven (1994 a). Living systems theory completed with break-through thinking in a municipal restructuring and planning process. *Systems Research and Information Science,* vol. 6.

Rasegård, Sven (1994 b). Information and society – an overview of certain overlapping problems. *Systems Research and Information Science*, vol. 6.

Rasegård, Sven (1996 a). A system syndrome – the crisis of a social welfare system. *Systems Research and Information Science*, vol. 7.

Rasegård, Sven (1996 b). Some aspects of the concept of behavior on different levels of societal functions. *Systems Research and Information Science*, vol. 6.

Rasegård, Sven (1997 a). A system Syndrome – the crisis of a social welfare system. *Systems Research and Information Science*, vol. 7.

Rasegård, Sven (1997b). Science and society – the need of promoting interdisciplinary research. *Systems Research and Information Science*, vol.7.

Rasegård, Sven (1998). Some notes on the system concept. *Systems Research and Information Science,* vol. 8.

Rasegård, Sven (1999 a). Some aspects of the usefulness of living systems theory in a municipal organization. *Systems Research and Behavioral Science*, vol. 16.

Rasegård, Sven (1999 b). Some notes on certain common problems in scientific research work. *Systems Research and Information Systems*, vol. 9.

Rose, Steven and Appignanesi, Lisa (1986).*Science and beyond.* Oxford: The Institute of Contempary Arts.

Roos, J.P. (1973). *Welfare theory and social policy.* Helsinki: Societas Scientarum Fennica.

Samuelson, Kjell (1978:1). *Informatics with systems science and cybernetics – concepts and definitions*: Royal Institute of Technology and Stockholm University.

Samuelson, Kjell (1978:2). *Informatics by general systems and cybernetics. Stockholm*: Royal Institute of Technology and Stockholm University.

Samuelson, Kjell (1982). *Systemvetenskap internationellt* (Systems science at certain universities abroad).

Schoederbek, Peter A., Kefalas, Asterios and Schoderbek, Charles G. (1975). *Managements systems: Conceptual Considerations.* Bussiness Publication Inc, Dallas.

Senge, P.M. (1990). *The fifth discipline.* New York: Doubleday/Currency.

Sinderman, Carl J. (1987). *Survival strategies for new scientists.* New York: Plenum Press.

Stockholms kammarkontor, *Stockholms stads årsredovisning* 1986 (annual report of the City of Stockholm,1986).

Svenska Dagbladet 28–29 June and 14th October 1995. Stockholm.

Taormina, Robert (1991). Organizational analysis of information processing using living systems theory. *Behavioral Science*, vol. 36.

Weinberg, G.N. (1975). *An introduction to general Systems thinking.* Wiley, New York.

Wiener, Norbert (1948). *Cybernetics.* Wiley, New York.

Ustinov, Peter (1995). *Legal systems where the only people who win are the lawyers.* London: The European 3-9 August, nr 273.

ABOUT THE AUTHOR

The author is a member of the Society of Living Systems. The Society of Living Systems is an association of free researchers within the interdisciplinary field of informatics and systems science. This association is a forum for exchanging information between researchers of different experiences and backgrounds (technology, economy, social science, music and others). One can say that the overlapping aim can be expressed by the following metaphor from a founder of the association, Mr. Tommy Kruszelnicki: " Whereas the sectional researchers work like industrious farmers in their field in the valley bound to a system, we strive like mountain climbers up to the heights with the aim to be able to see the entire beauty of all systems."

The Society of Living Systems is engaged in a research group named the Systemology Group. This group arranges seminars and produces scientific research reports published in the Systemology Journal and other journals within the wide area of informatics and systems science.

INDEX

A

B

D

G

H

L

M

N

O

P

Q

R

S

T

X

Y

Z